"Deconstruction is rather easier and more common than construction. In *Dimensions of Faith*, Dr. Donaldson takes the harder path and delivers a constructive work providing a hopeful voice to those of faith (all of us) by challenging current paradigms and imploring exploration into further understanding, reminding us that to arrive is to fall short."

—Todd Harrington, Lead Pastor, Haven Field Community Church, South Vestavia Hills, Alabama

"*Dimensions of Faith* tackles complex questions about faith from the viewpoint of a scientist who is also a Christian. But first, the author neatly unpacks assumptions, mindsets, and misconceptions about faith, and defines faith in a careful and logical way. He then establishes and explores the inextricable link between faith and reason, and analyzes, in depth, the nature of religious faith. This excellent book is a must-read for anyone who has seriously pondered faith and its role in our lives."

—Sharon Stuart, Lawyer

Dimensions of Faith

Dimensions of Faith

Understanding Faith through the Lens of Science and Religion

Steve Donaldson

CASCADE *Books* • Eugene, Oregon

DIMENSIONS OF FAITH
Understanding Faith through the Lens of Science and Religion

Copyright © 2015 Steve Donaldson. All rights reserved. Except for brief quotations in critical publications or reviews, no part of this book may be reproduced in any manner without prior written permission from the publisher. Write: Permissions. Wipf and Stock Publishers, 199 W. 8th Ave., Suite 3, Eugene, OR 97401.

Cascade Books
An Imprint of Wipf and Stock Publishers
199 W. 8th Ave., Suite 3
Eugene, OR 97401

www.wipfandstock.com

ISBN 13: 978-1-4982-2005-7

Cataloguing-in-Publication Data

Donaldson, Steve

Dimensions of faith : understanding faith through the lens of science and religion / Steve Donaldson.

xvi + 282 p. ; 23 cm. Includes bibliographical references.

ISBN 13: 978-1-4982-2005-7

1. Faith and reason. 2. Religion and science. I. Title.

BL240.3 D65 2015

Manufactured in the U.S.A.   07/27/2015

All Scripture quotations, unless otherwise indicated, are taken from the Holy Bible, New International Version®, NIV®. Copyright ©1973, 1978, 1984, 2011 by Biblica, Inc.™ Used by permission of Zondervan. All rights reserved worldwide. www.zondervan.com The "NIV" and "New International Version" are trademarks registered in the United States Patent and Trademark Office by Biblica, Inc.™

"Revised Standard Version of the Bible, copyright 1952 [2nd edition, 1971] by the Division of Christian Education of the National Council of the Churches of Christ in the United States of America. Used by permission. All rights reserved."

This book is dedicated to my wife Carol and our children Rebecca, Joshua, Matthew, Rachel, and Mary Katherine.

Contents

Acknowledgments | ix
Introduction: What Qualifies One to Build a Cairn? | xi

PART I—The Nature of Faith
    1 Saving the Assumptions | 3
    2 Who Believes What?—The Range of Human Beliefs | 7
    3 Misconceptions About Faith: What Faith Is Not | 13
    4 What Faith Is: Making Faith Concrete | 33
    5 Faith and Brains | 55

PART II—Faith and Reason
    6 The End of Knowledge: A Challenge for Faith | 79
    7 Crafting Rational Faith | 113
    8 Explorers and Mechanics: Creative Approaches to Faith | 139

PART III—Religious Faith
    9 The Nature of Religious Belief | 171
    10 What is Religion, Anyway? | 192
    11 How Clear Is God? | 201
    12 Is This Really What Religious People Mean by Faith? | 223

PART IV—Conclusion
    13 No End in Sight | 239

References | 251
Subject Index | 263
Scripture Index | 281

Acknowledgments

A BOOK WRITTEN IN isolation is probably best read only by the author. Despite any flaws of my own doing, this book does not suffer from that limitation. I am grateful to Frank Donaldson, George Keller, and Tom Woolley for reading and commenting on early drafts, and to students and colleagues for their insights and interaction. My wife and kids are to be commended and thanked for their patience during the many hours I spent working on the manuscript and for being the subject for many of the examples. Finally, I want to thank my parents, Patti and Frank Donaldson, for raising me with an insatiable desire to question, study, and learn. This book is a tribute to their efforts.

Introduction

Dimensions of Faith

What Qualifies One to Build a Cairn?

STUMBLING TO THE TOP of yet another Rocky Mountain peak, I discover that the cairn-builders have been hard at work. Ironically, the way is anything but clear. Although the function of a cairn[1] is to show the best path, at the moment there appear to be four or five suggested summit routes, each marked with its own metamorphic or igneous signage. At least the cairns announce that someone has been this or that way. But what kind of someone might that have been? These piles of stones indicate that lots of folks have had lots of ideas about the best way to reach the top of this particular mountain. As it is improbable that each way is equally good, I consider it more likely that the primary function of these cairns is to serve as tribute to the egos of their creators. If true, they are little more than the idols of those worshipping their own perceptions. Apparently, some of the cairn builders are not to be trusted, but which ones?

The one clear thing is that this proliferation of cairns has made my job harder. The paradox pleases me, however, for it means that I will actually have the chance to make some navigation decisions on my own. It hasn't always been so. On earlier trips, unsure of myself, lacking the explorer's instincts, or simply looking for the easiest way, I often found myself craving a cairn and not too picky about it. Sometimes, that was a mistake. Even a single cairn can be wrong.

Whatever their guise, cairns have their place, of course—most of us would not wish to guess which medicine bottle to open or which religion offers the best chance for immortality—but it is also apparent that a world with a cairn at every choice would be a bland place, indeed. More likely, it

---

1. A pile of rocks used to mark a route.

would be no better than a world with no cairns at all. One would still be forced to decide which cairns, if any, are valid. An endless flow of clear-cut answers seems even worse, leading quickly to erosion of the soul because the ideas and conceptions necessary for its stability have no chance to take root and grow. Searching, therefore, is more than an honorable profession—it is an essential component of development. Forced to evaluate the various alternatives for proceeding upward, perhaps even trying a few, we sharpen our way-finding skills and build endurance for further adventures.

Thus, although my current position on this mountain is testimony to the usefulness of cairns (in this case some reasonably accurate maps and prescriptions for proper training), it is also evident that too many—or the wrong kind—would actually be detrimental to further progress. Perhaps this is one reason why our requests for an endless supply of answers for life's difficult questions are not readily met. This need not imply that questions are more important than answers, but it does suggest that we take care in their framing. Even in cosmic *Jeopardy*, an answer only has value if someone is asking the right question. Furthermore, acknowledging that cairns can only take one so far is actually a step in the right direction because it precludes the accompanying insinuation that, by not providing the clear-cut answers we desire, God is somehow derelict in his job. There is no need to be towed by a master worth following. Indeed, it seems unlikely that such a master would ever tow the apathetic, much less the unwilling. Cairns are not some magic carpet—we still have to climb.

The implicit assumption is that, given the proper cairns, we would always climb in the right direction. This, however, is a fallacy. In the first place, humankind has a less than admirable track record of doing the right thing, even when the right thing is clearly evident. Nor does this just happen periodically on a global or national scale. Daily desires for immediate gratification frequently lead to poor choices, despite the evidence of a better way. It is not that the long term consequences are not clearly marked—they are merely ignored while we try a shortcut to happiness. Furthermore, guarantees of never making the wrong choice can only be offered to puppets. Under such assurances, the cairns themselves (or perhaps their builders) would be the puppet masters. Free will cannot exist in such a world.

So, clamoring for more cairns is often the mark of laziness, whereas claiming that one would always follow the right cairn is a sure sign of arrogance. Failure to evaluate and compare may be worse, for by another name it is known as blind faith. Far from being something one should proudly cherish, blind faith is little more than hope without substance, belief without evidence, and, ultimately, wishful thinking. For those with blind faith, any old cairn will do.

INTRODUCTION

Why Faith?

This book, you might deduce, is a cairn of sorts and I am thereby reminded of the admonition by mountaineer Gerry Roach to "trust your judgment, not some stranger's pile of rocks."[2] For most of those beginning this book, I am just that—a stranger—and whether I am one who can be trusted is a question worth asking. However, as with any cairn, one can only assess its ultimate viability and the trustworthiness of its creator after either reaching the goal to which it points or finding that it is poorly placed. If, in the process of reading this book, you discover new and productive ways to think about faith, take a hard look at the reasons for specific beliefs, and evaluate your own approach to truth, you will have reached the goals I intended to mark. Whether this book contributes positively to your journey neither of us can yet say but your decision to read it (or not) is an expression of your faith about what you believe are the likely consequences of doing so. It can be no other way because faith is woven into the very fabric of our being.

Unfortunately, many religious people believe that they have a corner on the use of faith and the non-religious are more than willing to agree. While the former see faith as some mystical but admirable quality in and of itself, the latter are often led to disparage it. Perspectives such as that of artificial intelligence pioneer Marvin Minsky that, "faith and critical thinking are mutually exclusive"[3] are typical. I think it is clear that Minsky has religious faith specifically in mind but what would he say of faith in all of its other manifestations? Although his statement may apply to some types of faith some of the time, in no way can it be construed as an accurate comment on the nature and role of faith in general. In fact, as I'll be careful to show, critical thinking itself is impossible without faith.[4]

This tendency to conflate faith with religion is understandable in part but it can cause one to dismiss the pervasiveness of faith outside that domain. It also predisposes us toward an unwarranted compartmentalization of cognitive function and can even lead to some rather silly proclamations such as that of well-known atheist Richard Dawkins that, "Atheists do not have faith."[5] Faith, however, is central to everything we are and do and, as Dawkins should know, is integrated into our biological natures.

---

2. Roach, *Colorado's Fourteeners*, 200.
3. Kuhn, *Closer to Truth*, 212.
4. Or, as Alister McGrath puts it, "Faith is part of the human condition. It is impossible to construct an argument proving the legitimacy of reason without presupposing faith" (*Surprised by Meaning*, 113).
5. Dawkins, *God Delusion*, 51.

Some of our beliefs are clearly of little concern but others have life-changing consequences. Deciding which is which is itself a matter of faith and, as a result, it makes more sense to understand where faith comes into play and how it operates than it does to deny its existence or downplay its role. In light of this, an obvious goal is to have the most accurate set of beliefs possible, particularly in those domains deemed important. Faith, then, is central to an ongoing search for truth but even the results of such quests must always be cast in terms of faith.

Consequently, faith is what we think is true about the universe and everything in it, including what we think is or is not in it. Faith is deciding how we will live our lives—deciding if we will focus on this or that, whom we will or will not trust, where our energies should be spent, what is and is not in our best interests and the best interests of those we care about, who it is that we should care about, what is worthwhile and what is worthless. Faith colors life and makes it possible. Everyone has faith in everything but the degrees vary widely among people and even within a person from time to time. Where and to what extent our faith is placed determines who we are and what we do. Some lives are colored primarily true and some primarily false, but all are colored both partly true and partly false, for that is the artistry of faith.

Vistas, Visions, and Voyages

Among the many advantages of climbing a mountain is the perspective it affords the climber—one that is unavailable to the person content to remain at its base. Several years ago when my three older children (at that time quite young!) and I hiked up Mounts Democrat, Lincoln, and Bross in the Mosquito Range of Colorado, we began our trek near a small body of water called Kite Lake. I gave little thought to why it might have that particular name until we had gained sufficient altitude for it to be readily apparent that the outline of the lake bore a distinct resemblance to the shape of a kite. In retrospect, I might have guessed as much but, if reduced to guessing, it would have been equally valid to imagine that the lake was named for someone named Kite, that it commemorated the experience of some pioneer who found the mountain gusts ideal for flying a kite, and so forth. In fact, in the absence of any disconfirming evidence, those would have remained possible explanations except, having seen the lake from above, they hardly seemed likely.

In general, dimensions that are all but unobservable from one perspective become visible with effort, possible explanations sort themselves out, and views of the truth become clearer. Regrettably, misapprehension

INTRODUCTION

of faith can destine us to a low dimensional world where it is all too easy to end up functioning with less than valid beliefs to which we cling with unbecoming passion. Not the least of those may be our view of faith itself. To debunk such anemic perspectives of faith, it helps to see its true multi-dimensional character.

As a college student studying physics I was aware of the potential conflicts between my religious faith and what I saw revealed through science but I have never been satisfied with the idea that the domains must be so separate as frequently proclaimed. Yet if faith operates on both sides of the supposed divide, then understanding faith becomes a reasonable starting point for any attempt at resolution. Most people may not be focused on bridging that particular gap but erroneous and unhealthy attempts by both religious and non-religious people to pigeonhole their lives into faith and non-faith parts create their own set of gaps—traps might be more accurate—of which many are conveniently unaware. It is also the case that poor conceptions of faith are not restricted to those with impoverished academic backgrounds because very well-educated and otherwise intelligent individuals frequently demonstrate naïve views of faith regarding both themselves and others.

My aim in this book, then, is to explicate the nature and role of faith and to ponder the implications of its proper conception for how we live. Part I considers the nature of faith. In chapter 1 we will contemplate how everything we think, believe, or think we believe is founded on assumptions that are often so far removed from our present state that we have lost sight of their origins. Nevertheless, we have a strong sense that saving those assumptions is paramount to maintaining our current beliefs and in that we are correct. Whether they should be saved is an altogether different matter, but, because some mindsets are nothing more than set minds, they may be all but impossible to change. In chapter 2 we'll ponder briefly why we are disposed to believe certain things as a prelude to considering the fact that the content of our faith differs from each other in dramatic ways. Naturally, we should want to know why that is the case. Chapter 3 deals with a number of common misconceptions about faith. Before we say what faith is we will try to understand what it is not. This leads us, in chapter 4, to a definition of faith that makes it easy to see how faith permeates our existence and is operative in virtually everything we do. Chapter 5 concludes the first part by showing that there are definitive reasons why we operate on the basis of faith—it is the natural and inevitable product of brains.

Part II examines the relationship between faith and reason. Chapter 6 begins this process by probing limitations that necessitate faith and explaining why faith will, therefore, always be necessary. Chapter 7 shows why any perceived dichotomy between faith and reason is a false one and considers how to move in the direction of rational faith. Given that we have no choice but to live with faith, getting things right is the name of the game. Consequently, chapter 8 develops a metaphor for thinking about how we approach the acquisition of valid beliefs and what it means to adopt one persona versus another.

Part III is devoted to an exploration of the nature of religious faith. Why do people think traditional religious faith is different from other kinds? Chapter 9 asks how people see religion and inquires into the nature of religious belief. This leads us to consider in chapter 10 what constitutes a religion and to ask if even science might not be a religion of sorts. Because a major sticking point for many individuals centers on God's perceived lack of clarity, chapter 11 asks, "How clear is God?" and reflects on whether he is doing his job well enough to warrant religious faith. In chapter 12 I consider various objections to my characterization of faith in religious contexts and indicate how traditional views of religious faith can be evaluated in light of other themes in the book. The final part (IV, chapter 13) attempts to imagine a world without faith and contemplates how truth seekers can sometimes end up in very different places.

PART I

The Nature of Faith

──── CHAPTER 1 ────

Saving the Assumptions

Whenever we couldn't conceive of what's out there, whenever we couldn't even begin to guess . . . it was because we didn't yet understand what the preconceptions might be that were restricting our view.[1]

—RICHARD PANEK

IN THE DAYS BEFORE cigarette advertisements were outlawed on billboards and television, it was not unusual to see an otherwise attractive man or woman staring at you with a prominent blackened eye and proclaiming (on behalf of a particular brand), "I'd rather fight than switch." Instead of touting the specific merits of that particular brand, the company had apparently decided to play to the American willingness to defend cherished and (presumably) worthwhile positions at any cost. It is not, of course, a mindset restricted to Americans but, as the advertisers surmised, it does go well beyond the choice of cigarettes. For example, consider a letter addressed to the editor of a major daily newspaper:

> Besides being an insult to man's intelligence, the concept of faith is dishonest and an intellectual copout. It lacks merit and credibility because it is entirely subjective and has no limits and no method of acquiring knowledge or distinguishing the difference between what's true and what's false . . . Reason, on the other hand, is a tool of critical and objective thought . . .[2]

1. Panek, *Seeing and Believing*, 175.
2. Bustamante, "Concept of Faith".

A day later, the following response was printed, submitted by another reader:

> The writer of the letter, "Concept of faith an intellectual copout" missed the point entirely. The beauty of faith is that by definition, it is subjective, voluntary, and not based on proof. Otherwise it would not be faith . . . The (other) letter writer is trying to deal with an abstract idea in his concrete and objective world.[3]

Both authors, it seems, would rather fight than switch but I suspect that their dogmatically expressed positions are not unique and that most of us would come down fairly strongly on the side of one or the other. Both views contain elements of truth but, anytime this occurs, we tend to see only those components that support our current beliefs and to ignore the other.[4] However, the apparently antagonistic positions they have adopted may obscure the fact that their letters share two important points. In the first place, both of these are statements of belief, exhibiting the faith of their respective authors on the nature of faith itself. Secondly, although these letters reflect widely shared but apparently opposing views on faith, those views are both fundamentally wrong.

Not only does faith *not* lack merit but, as I will demonstrate throughout this book, it is at the core of everything we think and do and is by no means limited to religion. Bad press aside, faith must always play a critical role, even (and perhaps especially!) in an increasingly technologically oriented society. While pitting faith against reason is an old gambit, it doesn't stand up, even to reason. Because of this, the subjective nature of faith claimed by both authors is, at best, problematic. The suggestion that objectivity plays little role in matters of faith ignores the fact that, often, quite the opposite is the case. Furthermore, the extent to which faith is voluntary requires significant clarification. As I will show, faith is a natural and necessary product of human brains and some aspects will frequently be beyond our control. In short, faith is not abstract and certainly not just an idea. As nineteenth-century wit Josh Billings once said, "I honestly believe it is better to know nothing than to know what ain't so."[5] It's hard not to agree with him but, as he acknowledges, that, too, is a matter of belief.

In any event, you might argue that the two letter-writers are not discussing faith in general but have specifically targeted religious faith. Although there is no way to support that position from the excerpts I have provided, I don't disagree but submit that such arguments are built on a gross

---

3. Eiland, "Faith Not Based on Proof."
4. Cf. Mlodinow, *Drunkard's Walk*.
5. Billings, *Everybody's Friend*, 286.

misunderstanding of faith itself. Because each view reflects a wide consensus of opinion, correcting that confusion is the first step toward moving us into position to better appreciate opposing viewpoints as well as our own.

This confusion often leads to an interesting irony. Ridiculing faith as scientifically deficient, a crutch for the rationally infirm, and generally fit for those who have no better basis for their opinions reflects the types of viewpoints one expects to hear espoused by those who have little regard for religion. But it is not just the non-religious who take a dim view of faith. Religious people often see it as a necessary evil—as a poor excuse for knowing and something that will, at some future time, no longer be needed[6] (i.e., "faith is necessary now but one day we'll have sight").[7] But for any religion that postulates a deity with infinite attributes, this, too, must be wrong.

None of this is to acknowledge that all faith is created equal—quite the contrary. Sometimes faith is the bad kid on the block—a weak-minded bully without credibility. Occasionally faith *is* rationally infirm and based on unsupportable assumptions. Although there is nothing fundamentally wrong with faith itself, there may be plenty wrong with any particular belief. In fact, faith in any given entity or idea can be weak or strong, unsupportable or well-grounded, and unthinking or highly critical. It would be a mistake, however, to believe that some of these terms are merely synonyms for others. For example, a faith might be strong but unsupportable, never having been subjected to critical reflection of any sort. The power of such faith can be great (i.e., in the sense of causing people to engage in significant activities) even when the evidence for the objects of that faith is flimsy, but that does not validate the faith. "Holy wars" have been fought between parties with strong faiths in their religious perspectives or scientific theories but it is not always possible for both to be right. On the flip side, critical examination will not necessarily render a faith powerless, although it can. The fact that it might is a scary proposition for many and reason enough for them to reject any tendency toward examination.

At Arches National Park in Utah, a large boulder perches precariously atop a slender rock pedestal beside one of the many turn-outs (see Figure 1). A prominent sign accurately but unimaginatively designates this "The Balanced Rock." As one strolls around the park one sees evidence everywhere of what must, surely, have been many previously balanced rocks, although none of those are labeled. The fear that cherished beliefs may similarly be subject to rapid and cataclysmic erosion—that a supposed rock-solid faith can be undermined and come tumbling down if one waffles on any part—is potent.

---

6. Or what Giberson refers to as "the burden of faith" (*Saving Darwin*, 164).
7. Cf. 1 Cor 13:12.

We're accustomed to being wrong but not to admitting it. As a result, we often go to great lengths to save our assumptions, even to the extent of being willing to fight, rather than switch. The black eyes accompanying such metaphorical (and sometimes literal) battles, however, are a poor reason for pride.

Figure 1. Apparently rock solid faith can sometimes rest on a precarious foundation.

It is helpful in such circumstances to note that the opposite of balanced doesn't always mean unbalanced—in our example it merely means stable. All those fallen (formerly balanced) rocks are no longer precariously supported but now rest on a firm foundation. Balance is no longer an issue. Like the balanced rock, weakly supported faiths call attention to themselves but are in constant jeopardy of toppling. Well-grounded faiths aren't always so flamboyant but they are the only kind with lasting value. Consequently, to be valid, faith must always be accompanied by critical examination. The process can be traumatic but it can also be stimulating and life-changing.

CHAPTER 2

Who Believes What?—The Range of Human Beliefs

> God can make a cow out of a tree, but has He ever done so? Therefore show some reason why a thing is so, or cease to hold that it is so.[1]
>
> —WILLIAM OF CONCHES

IMAGINE THAT WE HAVE just met at a party. During the introductions and obligatory small talk, I drop one of the following bits of information that, I claim, describes an actual event in my life:

1. I fell 20,000 feet without a parachute.
2. I once did 3,000 non-stop push-ups.
3. I landed in a tree in my kayak.

Which would you believe to be true? Having actually presented these options to others and noted their responses, I believe that I know which you are most likely to select. Of course, you don't have their advantage, so I might be wrong. For example, if my proclamation was made while you stared at my horribly disfigured body, you might be inclined to accept the first option as the most plausible. Alternatively, observing a tight shirt that could barely contain my bulging shoulders, you might lean toward option two. I'm not sure what visage I would have to present to make you think

---

1. Thorndike, *History of Magic*, volume II, book IV, chapter 37, 58 (referring to 1531 edito princeps of Basel, 29).

option three the most probable but I'm willing to bet that, in the absence of any additional information, it is the one you will select. (For the record, that would be the correct choice but I'll wait until the next chapter to try to convince you to believe me.)

Your actual choice, however, is not really the issue. What you should be asking yourself is why you made the choice you did, whatever it was. As you ponder that, you might also wonder whether you would have been more or less likely to believe any one of these statements if they were made about someone else rather than about me—for example, if I told you that I just read that *someone* fell 20,000 feet without a parachute. Claims one and two are both, in fact, presumably true about someone else.[2]

As for the reasons for your choice, there could be several. Clearly, the odds of any one of these statements applying to someone are better than for a particular someone (i.e., me). Or, maybe you are familiar with Guinness (the record book, not the beverage) and have run across enough unlikely stories that you give me (or someone) the benefit of the doubt. If you're a knowledgeable pragmatist, you may know how incredibly difficult it is to perform more than even 100 consecutive push-ups and rule that out while conjuring up ways a body falling at a terminal velocity of 120 or so miles per hour might not be smashed to smithereens. Perhaps option three seems a bit bizarre but mundane enough to be true, especially if you're not into believing things that seem incredulous (even though someone could have met you at a party, made statements one or two, and been telling the truth). And so forth.

In any event, your choice is a statement about your beliefs and is based on your background including your personal experiences as well as those stories you have determined to believe (or not). Similarly, my conjecture above was really nothing more than a declaration of my belief about your belief. Belief is what we do.

This is not to say that we believe the same things, and therein lies a wondrous phenomenon of the human condition. Although it might have been otherwise, it is hard for us to imagine a world or even a (non-trivial) scenario in which we would all share identical beliefs. Furthermore, the range of things that people believe is truly astounding.

One need not look far to discover people accepting assorted end-of-the-world prophecies, the efficacy of séances, or various paranormal processes. But the scope of belief is not restricted to views of the minority. Select practically any period of human history and you'll find large numbers

---

2. Ruffin recounts several cases of lengthy but survived falls (*Aviation's Most Wanted*, 257–58). The website http://www.recordholders.org/en/list/pushups.html lists the record for consecutive push-ups as more than 10,000.

of people subscribing to a belief in many gods, another group to belief in a single god, and still others with belief in no god at all. Surely they can't all be right. Moreover, beliefs in different eras about origins, the structure of matter, and the processes of life indicate that science is not immune, even if one does sometimes find a higher degree of consensus. Polls, such as those conducted periodically by organizations like Pew and Gallup, not only highlight the diversity of beliefs, they also chart their changes with time. But temporal variation in summary values only occurs when the beliefs of individuals are changing, and ours sometimes do. A dramatic shift in belief such as that by formerly prominent atheist Antony Flew[3] may be well publicized but most of us can point to noteworthy shifts in our own lives, possibly equally striking.

So, this is personal. Beliefs are at the core of who we are, both as individuals and a society. They place us in conflict with some and in harmony with others. Sometimes the turmoil is strictly internal. On occasion the conflict can be heated. At times it seems inconsequential. A teenage boy believes he looks suave with unruly hair that reaches his shoulders while his mom believes he is far more handsome with shorter, well-groomed locks. The Marines believe differently still. Trivial beliefs, perhaps, but important to those who hold them and easily elevated to the non-trivial if the issue changes from removal of one's hair to removal of a cerebral tumor where the risk of surgery is believed (by one physician) to outweigh the risk of non-intervention (as suggested by another).

If you're paying close attention, you may object that I have been a bit loose with my choice of words. Why say the boy, mother, Marines, or physicians "believe" this way or that? Why not just acknowledge that this is what they "think." Why not admit that each has different objectives and that each perspective is really just a matter of context? Ok, I admit it. But this kind of thinking *is* their belief. Although not all thinking pertains to belief, the two are often so intertwined with our perspective as to be synonymous and context is always a factor. This is not to say that all beliefs are equally thoughtful—examining why people entertain certain beliefs is often more illuminating than merely observing what they believe.[4] Their reasons can vary widely, from complex analysis to wishful thinking and, frequently, to almost no thinking at all. Soon we will explore this in much more detail but for now it is sufficient to note that belief permeates all of our thinking, from the deep questions of meaning and value to the way we imagine we look. In

---

3. Flew and Varghese, *There is a God*.
4. Cf. Gilovich, *How We Know*.

fact, just acknowledging that all perspectives are really matters of belief can precipitate change in some of those very perspectives.

Furthermore, beliefs are contagious. On any given autumn Saturday in the United States, significant numbers of individuals with great differences in their beliefs as to which team should be favored by the sports gods can be found occupying opposite sides of rather large stadiums, each group infected after a multi-year exposure to a particular collegiate climate. Inconsequential? Maybe, but such a belief contagion is clearly mirrored at larger scales where the objects of one's faith—in the political arena, perhaps—can make critical differences for an entire nation. Interestingly, we sometimes have an adverse reaction to the beliefs to which we are exposed and end up on what appears to be the opposite side of the belief fence from the progenitors of those beliefs—teenage rebellion is an obvious, if not dramatic, example. As we'll later see, however, some of these clashes are more apparent than real.

Observation of conflicts in belief suggest that the venues for our faith come in opposites—that you and I will either agree or have diametrically opposed beliefs about a thing. One of us believes in free will but the other does not. There either is an immortal soul or there is not. Human existence has an intrinsic purpose or it does not. And so forth . . . Yet, in most cases, there is actually a range of beliefs related to a given topic and the polar opposites merely receive most of the attention. Richard Dawkins, for example, classifies himself as a six on his own one to seven scale where one denotes complete assurance that God exists and seven denotes complete assurance that he does not.[5] We could assign the belief about the merits of a one to seven scale to another one to seven scale but only if the beliefs it reflects cover the entire spectrum of possibilities.

Recognizing the range of beliefs is not surprising but it should make us pause to consider why beliefs can vary to such a significant extent on matters both important and not. The fundamental cognitive mechanisms that underlie belief formation are, after all, shared by everyone. Although there are minute differences in the initial wiring of each brain, it seems apparent that the predominant cause of differing beliefs is simply related to differences in background.[6] In the ongoing nature/nurture debates, aside from the largely discredited ideas of Lamarck,[7] this has seldom been questioned, as few could conceive how anyone could be born with a ready-made set of

---

5. Dawkins, *God Delusion*, 50–51.

6. Small initial differences in brains can also have large effects but that is the exception, not the rule.

7. Lamarck proposed that traits acquired through learning or effort could be inherited. Except for the Baldwin effect and the possibility of certain epigenetic influences, his ideas are not currently accepted.

beliefs.[8] Even if there are genetic tendencies to certain types of belief—postulated, for example, based on observations that humans appear to share an innate religious instinct—specific beliefs are always developed in some cultural context.

But is it important to be on the same page with regard to our beliefs? Does it matter that we believe the same things? Isn't there value in diversity? It is not hard to imagine the disastrous consequences that would ensue if we all believed that we should pursue the same career or marry the same person. Clearly, life would be less interesting if we shared every belief, but it is also apparent that there are plenty of areas in which conflicting beliefs can have significant and potentially disastrous consequences. For example, you and I may differ in our beliefs about how best to care for the environment but unless we share the general belief that there is a need to do so, we will likely end up with one of two very different kinds of planet.[9]

It is not at the species level only, however, that differences in belief have significance. As noted earlier, belief is personal. From religious convictions to decisions pertaining to education, occupation, lifestyle, and finances, our beliefs affect us in important and lasting ways. In my collection of memorabilia is an old invitation to a wedding tea, given in honor of my wife prior to her marriage to someone else—a marriage that never took place because she called it off two weeks prior to the event. The engagement was only possible because both had faith that marriage was an appropriate next step in their relationship. Luckily for me, her beliefs could change (for which our kids will forever be grateful).

Now, if I were writing the script for one of the many television shows that parade people's petty pasts, I might speculate ad nauseam as to the reasons for her change in belief (for TV, we would label it a change of heart). But that is not our goal. Suffice it to say that it was not a matter of looks, physique, intelligence, job, hobbies, philandering, or fiscal irresponsibility that convinced her she was on the verge of making a mistake (my term, I admit). The bottom line—and this she admits—was that they didn't share certain basic beliefs. In other words, her belief about their relationship and, eventually, about our own, was not only based on her other beliefs, it could not exist without them.

This observation about the pervasive integration of beliefs into every aspect of our existence does not constitute some shocking new perspective on human thought, but I am suggesting that it is seldom granted the significance it deserves. Consequently, let's try one more example to drive this

---

8. Cf. Locke, *Human Understanding*.
9. Cf. Diamond, *Collapse*; Wilson, *Creation*.

point home. To do so, return to our environmental scenario and imagine that I am what someone would label an "insensitive ecological moron." In other words, I believe that I am perfectly justified in consuming resources willy-nilly, in discarding my trash wherever I please, in laughing at the very concept of clean energy, and so forth. What, pray tell, could explain my beliefs? Actually, there are any number of possible explanations, including an overestimation of the capacity of the earth to absorb my waste, miscalculation of the cumulative effects of dimwits like me, or some religious impulse suggesting that this world doesn't really matter. Regardless, the key idea is this: no belief exists in isolation. As in both examples, each and every belief is typically supported by (and in tension with) many other beliefs. We are belief machines, blending, mixing, and matching that special concoction that makes us who we are at any given moment of our lives. Thank goodness, our tastes can change.

Nevertheless, although it is painless enough to acknowledge that our beliefs *can* change and equally easy to think that the beliefs of others *should* change, it is far harder to admit that our own beliefs may actually be in need of revision. But some beliefs are inevitably wrong.[10] Surely we aren't so vain or deluded as to suggest that none of those are ours? In addition, because a number of errant beliefs will necessarily be about things with crucial and perhaps enduring consequences, it behooves us to get them right. An understanding of the nature and role of belief is essential to this process but, before we can continue our journey, we need to take a detour to consider several wrong-headed views of faith itself.

---

10. The fact that so many people can be wrong about the same things leads to entire books devoted to debunking false beliefs (e.g., Numbers, *Galileo Goes to Jail*; Burnham, *Dictionary of Misinformation*). One goal of this book is to counter false beliefs about faith.

CHAPTER 3

Misconceptions About Faith: What Faith Is Not

I beseech you ... think it possible you may be mistaken.[1]
—OLIVER CROMWELL

BEFORE YOU BEGAN READING this book, you already had some concept of faith. Furthermore, your act of reading indicates that you are willing to have that concept challenged (or at least that you are curious about my concept). But as I have already suggested, such a concept is itself really just a belief—in this case a belief about faith. We might call it a meta-faith. Meta-faith, in this case, would be one's belief about the nature and role of faith. It would describe what faith is, how it operates, and what influence it might or might not have.

Because meta-faith is itself a belief and because, as we've seen, some beliefs are wrong, it is inevitable that some meta-faiths will also be mistaken. As it turns out, there are a number of widely held views on faith that don't stand up to serious scrutiny. Many of these are justified inductively from examples selected only to make someone's preconceived point but, as any student learns in elementary logic, generalities are never proven by example. Consequently, even though there will always be situations in which faith appears to present a certain face, it is important to identify those situations for what they are and not to permit them to lead us astray (in the sense of making faulty generalizations). In this chapter, therefore, we will examine a

1. Carlyle, *Cromwell's Letters*, 448.

number of flawed beliefs regarding faith in the hope that, by clarifying what it is not, we will be in a better position to understand the true possibilities for the nature and role of faith. Below I have listed several prominent but mistaken views about faith. We'll dismantle these one by one.

Seven Common Misconceptions About Faith

1. Faith is not based on evidence.
2. Faith equals belief in the impossible.
3. Faith is antithetical to reason.
4. Faith is blind.
5. Faith is a substitute for sensory experience.
6. Faith is binary.
7. Faith is a religious phenomenon.

Misconception #1: Faith is Not Based on Evidence

Ambrose Bierce once described faith as "Belief without evidence in what is told by one who speaks without knowledge, of things without parallel."[2] Now, satirical statements are geared to make a point about the flaws in something, often with humor, and seldom with kindness. This take on faith by former cynic Bierce (former, only because he is now dead) is a good illustration and we can all point to instances where his assessment appears to ring true. I'm thinking, for instance, of the mob's attempt to burn a woman believed to be a witch in the comedy *Monty Python and the Holy Grail*.[3] The humor there, however, distorts the gravity of real executions of presumed witches in the early history of our country, which may appear to be sufficient reason to sympathize with Bierce regarding the incompatibility of faith and evidence. But the potential implications of one's beliefs, though frequently momentous for the believer or their peers, are not at stake here—evidence is. Furthermore, like most of the misconceptions we will discuss, this one is not limited to the anti-religious. Wendell Berry, for instance, thinks that, "religious faith *begins* with the discovery that there is no evidence,"[4] a

---

2. Bierce, *Devils Dictionary*, 40.
3. Gilliam and Jones, *Monty Python*.
4. Berry, *Life is a Miracle*, 28.

thesis that, if true, would seem to render any particular religion as good or bad as any other.[5]

The mistake of both Bierce and Berry is to claim that faith is not based on evidence. Actually—and this will surprise some folks—faith is always based on evidence. Sometimes, as in the case of "witches," the evidence is so thin as to be laughable but without some evidence there will be no faith. Faith can be good or bad in the sense of being valid or not but it is, in all circumstances, based on evidence. The evidence, of course, may be robust or flimsy.

The American judicial system presents a microcosmic view of the interaction between faith and evidence that is characteristic of that correspondence in all areas of our lives (except that we go further and permit various levels of hearsay to count as evidence, having noted that not all hearsay leads to heresy). Individual members of a jury form their own (often differing) decisions based on evidence of varying quality presented by the lawyers and received via testimony but that evidence is necessarily filtered and interpreted from the perspective of their personal experiences (which is, obviously, just a subsidiary form of evidence). Failure to reach a unanimous verdict is confirmation that the same evidence does not always result in equivalent faith. Nevertheless, no one would claim that faith in the guilt or innocence of the defendant was not based on evidence since it is clear that evidence played a role in either type of faith.

Simplistic categorizations of faith based on observations of unsupportable (or supportable) examples of specific beliefs is the induction fallacy and we recognize it clearly in other areas. No one, for example, believes all policemen are on-the-take just because some are periodically identified. Why should we think the concept of faith is corrupt just because some faith is tainted?

But, you might ask, can't robust faith be grounded on flimsy evidence? Don't some jurors make poor decisions about which they are supremely confident? Don't we all? Evidence must be interpreted, and what appears fragile to me may appear strong to you. Those things one person considers to be no evidence, others may find compelling.

Consider, for example, my earlier claim which, I submitted, was really true—that I once landed in a tree in my kayak. Should you believe me? I imagine that my choice of verb (i.e., "landed") may have placed you in a bit of a quandary because, although it is frequently associated with boats, it probably appears inappropriate for a context also involving trees. Perhaps

---

5. Something many atheists would readily affirm although, by adopting a position such as Berry's, how could one know?

you would be more comfortable if I had said that I once landed in a tree in an airplane. But that is untrue.

Now there are several kinds of evidence I could present to attempt to convince you to believe me about the boat. For instance, I might take a stab at explaining, via a bit of elementary physics, how it is possible to land a boat in a tree.[6] On the other hand, if you knew me, you might just chalk it up to my usual karma (although if you knew me you would know better than to call it karma). Or, my credibility might be reason enough for you to believe me (or not). Showing you the bent kayak would be nice. A photo of me in the boat in the tree would be better still. Any previous experience on your part, either in a kayak or observing others, would probably be helpful.

In the end, any of these reasons might be enough to convince you, but then again, none may be—evidence is not a synonym for proof. At this point you are having to take my word for everything anyway; although, if you had witnessed the event personally, you would probably think me stupid for trying to convince you at all. Yet, we're seldom fortunate enough to see most things firsthand. As I'll show later, however, even when we do, faith is still operative. When it comes to beliefs with more serious consequences, the need for cumulative evidence to make those beliefs robust and meaningful becomes paramount.

We can get another take on the evidence issue inherent in Bierce's contemptuous view of faith by taking his original statement and removing two of the prepositional phrases. This leaves us with the following: Faith is "belief . . . in what is told by one who speaks without knowledge . . ."[7] I hope that something strikes you as wrong about this. According to Bierce, the kinds of belief we would classify as faith are based on information obtained from those who don't know what they are talking about. This happens sometimes, certainly. But the logical equivalent is that if someone who does know what she is talking about tells us something, then it doesn't require faith to believe her. On the surface that, too, may give the impression of being reasonable. Can't we just accept what a knowledgeable person tells us? Why bring faith into it at all? But how can we be certain that someone knows what she is talking about? How do we assess credibility? Normally truthful individuals occasionally make mistakes. Habitual liars sometimes tell the truth. Consequently, we must actually make two decisions, one pertaining to the general credibility of the speaker and the other related to the content of her specific

---

6. The buoyant force on a kayak submerged deeply in a breaking wave can sometimes launch the boat into the air. (Thanks, Archimedes!) Whitewater kayakers call this an "ender." My landing was abruptly terminated by an ill-positioned forked tree, making the statement true.

7. Bierce, *Devils Dictionary*, 40.

message. Plus, those assessments can reflect a range of confidence values so that, whether we accept what someone tells us or not, faith is required on all counts. Faith, then, is at the very heart of credibility. But what if it pertains to unparalleled events?

Misconception #2: Faith Equals Belief in the Impossible

In the perspective presented above, Bierce also suggested that matters requiring faith are about "things without parallel."[8] Anytime something new transpires, faith is indeed required to accept it, particularly if we fail to participate directly. A number of people were incredulous when we went to the moon, split the atom, and observed experimentally that mass varies with velocity—all things without parallel when they first occurred. But then, so are an individual's first date, home run, taste of lemon pie, or heart attack. Yet, without firsthand experience, we can only accept (or not) what we learn from secondary sources through the gateway of faith. Until I walk on the moon, I must trust someone else's account of what it is like. But, things without parallel don't have to remain so. Moon landings are repeated, we help ourselves to another piece of pie, we hit another home run, and the initial events are no longer without parallel. Familiarity, however, not only breeds contempt, it spawns an increase in our confidence about subsequent outcomes. In other words, it increases our faith. At the same time, we are always left to wonder if, the next time, an event really will parallel previous ones.

So, faith never sleeps. Contrary to Bierce's claim, faith is always operative, even in the mundane and iterated areas of life—even in domains that are not without parallel. Maybe some marriage is radically different from all the rest but most are pretty normal. Yet faith is still required. We eat with the faith that our meal won't harm us but food poisoning is not uncommon. Passengers board planes with the confidence that they will arrive safely at their destination but, sometimes, they do not.

Some things, however, are truly without parallel. We may get only one chance at a heart attack. Virginity can be lost only once. There was only one Lincoln. Even first landings cannot be repeated. Bierce's sarcastic tone implies that faith is about believing the impossible but one-time events are without parallel in a unique sense and, until they happen and then forever thereafter, otherwise impossible. Faith is most definitely operative when we are confronted with them, but the need for such faith doesn't rule them out by fiat.

8. Ibid.

I hope it doesn't seem to you that I have taken this too far. I doubt that Bierce ever intended his statement to be digested and spit back to such an extent. I have done so, however, because this view is relatively common, and we can find enough examples where his appraisal rings true so as to begin to think that it is descriptive of the whole of faith. In reality all it does is describe faith's shallows. Deeper harbors are always available.

It is just that depth, however, for which many people who (sometimes unwittingly) disparage faith fail to account. It is a small step from the claim that faith is always about "things without parallel" that are related by "one who speaks without knowledge"[9] to the conclusion that people who tell about improbable occurrences are either misinformed or insane. Although history is full of improbable occurrences about which, if someone had not related them, we would never know, that dubious step was, nevertheless, taken by the philosopher Friedrich Nietzsche, who allegedly noted that "A casual stroll through the lunatic asylum shows that faith does not prove anything."[10]

We might attempt to discredit his view by noting that he ended up in an asylum himself, but doing so would only make us guilty of the same error he commits. Not only can crazy people occasionally make valid statements but things that sound crazy are sometimes true. Furthermore, if "faith does not prove anything," then Nietzsche's claim, which could only be made and can only be accepted on faith, is itself not provable based on his own assumption.

Naturally, where we stroll determines what we can observe. It would be a mistake to draw conclusions about the fertility of the North American continent based only on consideration of Death Valley or land inside the Arctic Circle. Similarly, we shouldn't expect a visit to an asylum to shed much light on cognitive features as they are operative in normal, healthy subjects. If we buy into Nietzsche's pseudo-logic, we might agree that a casual stroll through the annals of NASA disasters shows that faith does not prove anything whereas, by broadening our scope of observation, we discover that faith was a necessary precursor to virtually every NASA venture (even if a few did happen to end tragically).

We can almost always find ways to categorize something on the basis of unflattering examples and the poor use of induction. But derogatory comments never prove anything,[11] even though it is not unusual to hear

---

9. Ibid.

10. Attributed to Nietzsche in Winokur (*Portable Curmudgeon*, 97) without citation and repeated on countless web pages, this quotation appears to be a synopsis of what he actually said in Nietzsche (*The Anti-Christ*, 65).

11. I think it was my 12th grade English teacher who pointed out that name calling was one of the primary forms of fallacious reasoning. Obviously, Nietzsche missed that lesson.

them when something is deemed impossible. A popular tactic for implicitly engaging in such disparagement is to create a "straw man"—a weak and uncharacteristic example of whatever it is one wants to denigrate—and then to proceed to poke holes in that. Although easy enough to do, this is logically meaningless. Nevertheless, such an approach is relatively common, presumably either because the perpetrators will use any means available to make their points or because they are actually ignorant of the deeper aspects of their subject matter.[12] C. S. Lewis, for example, noted the tendency of some people to ask deep questions about the Christian faith but be unwilling to contemplate a mature response to their queries.[13] This is probably true of any religion but such shallowness is in no way limited to matters of religious faith. Even unpopular scientific theories can be turned into straw men by those who wish to discredit them.[14]

Misconception #3: Faith is Antithetical to Reason

When we use words such as "impossible," it is important to be perfectly clear about what we mean. Technically, this word denotes that which can never, ever happen under any circumstances, conceivable or not. Now, if faith equals belief in the impossible in this sense, then faith and reason are, indeed, incompatible. Faith would be solely for dummies and the hostility it sometimes breeds with reason, entirely justified. Clearly, if faith is just for the ignorant, reason should have nothing to fear. But this is a false dichotomy.

If you pay attention to how "impossible" is most often used, you will likely discover that what is really meant is, "This is what I believe can never occur." In other words, designating something as impossible is, itself, an act of faith (and this is true regardless of the sense in which the word "impossible" is intended). Thus, either way—whether we deem something technically impossible or merely improbable—belief is operational.

But what is the nature of that belief? This is the real issue. The infamous curmudgeon H. L. Mencken defined faith, "as an illogical belief in the occurrence of the improbable,"[15] but that can't be right. It suggests that belief in the improbable must either be illogical or must be called something other than faith. But what would you call it? Hopefully, if it really

---

12. A characteristic example is Kurtz, *Science and Religion*—lots of religious straw men.

13. Lewis, *Mere Christianity*, 36.

14. For example, consider some of the arguments employed in attempts to deny biological evolution.

15. Mencken, *Prejudices*, 267.

is illogical, you'll just call it illogical faith. In another universe Mencken's anti-twin might have defined faith as "a logical belief in the occurrence of the improbable" but this, too, would be wrong. Either position might be correct when describing a specific instance of faith but is inappropriate for portraying faith in its broader sense. In any case, belief in the occurrence of the probable need not be logical (because one can have all sorts of bad reasons for believing something that is true) and belief in the occurrence of the improbable need not be illogical (e.g., consider the claims of quantum theory for possible but low-probability events).

Unfortunately, the misconception that faith and reason are at odds is even enshrined in some formal descriptions of faith. For example, the first two definitions for faith offered by a popular dictionary are:

1. A confident belief in the truth, value, or trustworthiness of a person, idea, or thing.
2. Belief that does not rest on logical proof or material evidence.[16]

Now it is typical to use the terms faith and belief synonymously (and I will do so in this text) but merely defining one in terms of the other is of little help in understanding either. The real problem with these "definitions," however, is that they actually contradict each other at a fundamental level. Isn't it the case, for instance, that most genuinely confident beliefs "in the truth, value, or trustworthiness of a person, idea, or thing"[17] rest on evidence and hence logic? After all, prospective employers demand verifiable credentials, banks request identification before issuing funds, and we expect references before hiring a baby-sitter. But if that is faith (first definition), then how could it also *not* rest on logic or evidence (second definition)? One might also wonder how confident a belief must be before it can be called faith. Might we not sometimes proceed with doubts, even in the face of overwhelming evidence?

In reality, of course, these are less definitions of faith than they are indications of how some people use the term (e.g., Mencken could have written the second definition, confident that he possessed evidence to support his belief about faith). Although that might work for some words, in this case the ambiguous descriptions are inadequate as definitions. Thus faith may or may not rest on sufficient evidence and logic and it may or may not be confident, but we still need a suitable definition on which to proceed. I'll provide that in the next chapter but for now we can note that it is a misconception to merely declare it the antithesis of reason.

16. DeVinne, *American Heritage Dictionary*, 486.
17. Ibid.

Despite this, the idea that faith itself is not logical is pervasive and is not restricted to cynics. I once met a fellow who described his own religious belief as "illogical." Mencken would have approved but that description should not be mistaken for a comment about faith in general or religious faith in particular. To summarize, therefore, we can note that it is a simple matter to find examples of both irrational and unfounded beliefs as well as firmly-grounded ones. Some of these are in the commonplace and others in the improbable. People can (and do) believe anything and everything. So, even though faith can be either logical or illogical, it is still faith. An unreasonable faith, however, is nothing of which to be proud. Therefore, shame on my acquaintance! Because the interplay of faith and reason is critical, we'll devote considerable ink to this subject a bit later.

Misconception #4: Faith is Blind

When we understand that sound faith is always based on evidence, that faith need not equal belief in the impossible, and that faith and reason are not necessarily hostile to one another, it is relatively easy to see that faith need not be blind. Nevertheless, much like the previous misconceptions (and perhaps because it is so intimately related to them), this one is rampant. But how do people get this idea? Some faith, it seems, must appear to be blind, perhaps even to those who hold it. It is has frequently been presented as such.

Increase Mather, the mid-seventeenth to early eighteenth-century minister and past president of Harvard suggested, for instance, that people are kept in the dark so they "might live by faith" and "follow the Lord, as it were blindfold, whithersoever He shall lead them."[18] Former Salvation Army official Lyell Rader has suggested that "faith grows only in the dark."[19] It would be more accurate to say (regarding faith) that only blind faith grows in the dark. It is not true, however, that blind faith grows *only* in the dark. It is harder, but blind faith can also grow in the light if one closes one's eyes—that is, when failing to take note of or comprehend the evidence. In fact, it may be a disregard for or misunderstanding of the role of evidence that has given rise to this misconception. In a broader context,[20] Rader is

---

18. From his sermon "A Discourse Concerning the Uncertainty of the Times of Men." Miller, *American Puritans*, 190.

19. Attributed to Rader in Cory, *Quotable Quotations*, 129.

20. Ibid. The more complete quotation is, "Faith grows only in the dark. You've got to trust Him where you can't trace Him. That's faith. You just take Him at His Word, believe Him, and grip the nail-scarred hand a little tighter. And faith grows."

probably claiming that trusting someone on the basis of past experience can lead to an increase in one's faith (in that someone) as one continues to observe the expectations of such trust fulfilled (and I suspect Mather would agree).[21] However, one cannot make such a determination in the dark.

In any event, although faith itself is not necessarily blind, it sometimes appears to be so. Examples are not difficult to find. Several years ago I came across some pedestrians ambling down a two-lane road at dusk with their backs to oncoming traffic and wearing no reflective materials. Here, I thought, as I approached them in my car, is truly blind faith. Thomas Huxley labeled "blind faith the one unpardonable sin."[22] We might debate his numeric conclusion but it is tempting to agree that such faith is inexcusable. Yet, even though poorly supported faith (in many things) is, indeed, quite widespread, truly blind faith is all but impossible. In fact, it is hard to think of any faith that is not based on sight of something. Such vision may be 20/20 or myopic but each belief has its causes.

Even the apparently clueless pedestrians mentioned above no doubt elected to maintain their tentative position in the road because neither they nor anyone they knew had ever become a hood ornament under such circumstances. The consequent faith that whoever happened to be approaching them from behind was not himself blind, would see them in time, would move to the other lane, etc., might be flimsy but it was not without a perverse sort of rationale. I could have driven my point home but, as they anticipated, I swerved into the other lane, which helps explain why I am not making this observation from prison. Nevertheless, it is hard to imagine those pedestrians not moving to the sidewalk or at least into the other lane if they had given much thought to their circumstances. Hopefully it is clear by now that all faith is not this feeble. What we term blind faith is usually little more than wishful thinking—if, that is, any thought is involved at all.

We should also note that there is frequently a difference in what people say they believe and in what they actually believe. This is not to imply that when we do this we are outright liars (although that is a possibility) but that such disparity may reflect little cognitive engagement on our part. For example, it is not difficult to find people who say they interpret the Bible literally. Although I suppose there may be some, I have never met anyone who actually does. This is easy enough to see, for example, where despite the Apostle Paul's admonition against marriage, plenty of self-designated literalists go ahead and do it anyway.[23] Or consider the fact that, even though

---

21. I'll discuss this in more detail later when we specifically consider religious faith.
22. Huxley, "Improving Natural Knowledge."
23. Cf. 1 Cor 7.

Jesus said it would be better for someone to rip out his eye or cut off his hand than through them to fall for some temptation,[24] hospital emergency rooms garner little business from this practice. We could go on but differences between proclaimed and actual belief is not restricted to religion. Most of us need look no farther than the mirror to find an example. This is easily seen every time we claim to be capable of something we never actually try, excusing ourselves for various reasons when the real culprit is that we lack the requisite faith to do so. A bit of reflection would reveal the "faith" for what it really is.

It is just the types of uncritical thought we have been exploring that can lead to faulty generalizations from sound bites such as Bierce's, Mencken's, Rader's, or Huxley's. Mencken, for example, believed that, "The most costly of all follies is to believe passionately in the palpably not true. It is the chief occupation of mankind."[25] Judging from the number of websites that display this quotation (or a variant of it), this must reflect the sentiments of quite a few folks. But, even though it is hard to argue that such belief is not costly, it can hardly be as endemic as Mencken proclaimed or our society would fall apart. Indeed, were it our "chief occupation" we could not expect to survive very long, as any evolutionary biologist would be quick to point out. In order for a poorly conceived faith to endure, its immediate consequences must be minimal or the consequences must be postponed significantly. Hopefully, therefore, we are not tempted to mistake passionate belief for folly[26] or led to think that things which aren't true are always obvious.[27]

Misconception #5: Faith is (Just) a Substitute for Sensory Experience

The truth is, sensory experience is always obtained in some mental context and its interpretation is not infallible. The cognitive psychologist Daniel Schacter describes an incident in which a woman was raped by a man she subsequently identified as a prominent psychologist in the area. Fortunately

---

24. E.g., Matt 18:8–9.

25. Mencken, *Prejudices*, 616. I'll have more to say about the cost aspects of this later.

26. Compare, "Man is only truly great when he acts from the passions." Disraeli, *Coningsby*, 180. But, like faith, the merits of passion depend on context. Great deeds may be the results of passionate belief—it is hard to imagine them coming from either impassionate belief or passionate uncertainty—but so are many horrific or just plain foolish deeds.

27. Is it obvious, for instance, what constitutes the height of folly? Others have suggested something different (Cf. Ps 14:1).

for the accused, he had an airtight alibi, having appeared on a live television show at the same time as the crime. As it turned out, her TV was tuned to the station airing the show featuring the scientist and she had, apparently, superimposed his face on that of the assailant.[28] Our mistakes are seldom so dramatic but they occur, nonetheless.

My own experience with this illustration is a case in point. I originally encountered Schacter's description more than ten years before writing this account and wrote the first draft of this chapter under the mistaken belief that it was a physician (not a psychologist) who was involved. It didn't help that I had reinforced that belief by numerous retellings of the story (physician and all) to various classes through the intervening years. My belief was formed despite sensory experience (i.e., I read Schacter's account with my own eyes) and it was through the same type of sensory experience that I corrected my mistake before sending this to the publisher (i.e., I looked once again at Schacter's book) but it should be apparent that, in the future, I might recall the accused as being a physiologist, paleontologist, or proctologist or simply relapse into my original belief that he was a physician.

It would be easy to argue that our knowledge is accurate immediately after sensory experience (i.e., that I knew for at least a short time after reading Schacter's story that it was not about a physician) and that it is faith which must tide us over between sensory episodes. However, although it is true that our beliefs are apt to be more veridical the closer we are to some personal experience, as we will see in Chapter 5, even those beliefs are not always trustworthy.

In addition, the context in which the interpretation of sensory experience occurs is not restricted to our immediate circumstances. Take a look, for example, at Figure 1. What do you see? Look closely before reading on.

---

28. Schacter, *Searching for Memory*, 114.

MISCONCEPTIONS ABOUT FAITH: WHAT FAITH IS NOT

Figure 1. What do you see?

I've conducted this experiment on a number of occasions by actually displaying a similar slide to audiences and no one has ever denied seeing a mountain. The photograph is of Mount Moran in the Teton range in Wyoming and, once or twice, someone has actually recognized it. But I'm not primarily interested in their direct sensory experience or even yours. What I really want to know is what they (and you) perceive in the deeper cognitive sense of the word. What do you *see*?

This is where our backgrounds begin to take over. If you are an artist or a photographer, you might deem this striking mountain a worthy candidate for your creative endeavors, perhaps imagining how it will look at sunrise or partially obscured by clouds and assessing various possibilities for the foreground components of your masterpiece. If geology is your trade, you will immediately be drawn to the Black Dike (partially visible to the left of the upper glacier), a massive intrusion of igneous rock into the granite monolith, formed by some cataclysm occurring ages ago. Your interest in the mineral composition will, no doubt, be substantial. If you happened to be a prospector encountering Moran in the latter part of the nineteenth century, the only minerals likely to attract your interest would be gold and silver. Historians would see that, plus a great deal more, and could regale us with stories of pioneers at its base, early attempts on its summit, and the like. Climbers would see it as a peak to be scaled; skiers, as one to be descended. And each would see it differently before and after their respective adventures. Far from considering its recreation potential, early settlers might simply perceive it as a barrier to their westward progress. Had they been around, the Army Corps of Engineers would have been happy to remove it for them. Fishermen or water skiers may note the mountain but ponder their future on the river in the foreground.

Any of those perspectives, as we know, can change with time. When I look at Moran I see my younger son Matthew perched on its upper granite face, hundreds of feet above the Falling Ice Glacier and thousands of feet above the valley floor. I see the broken sandstone on its summit, indicative that it was once at the bottom of a substantial body of water. And I see the small ledge on which my son and I bivouacked before being able to complete our descent.[29] But it wasn't always so.

Thus, my perspective of Moran is different from virtually every other person's on earth (and even different from my own at an earlier time) but, then, so is yours. How could it be otherwise? Despite the fact that we could, at least theoretically, share the same sensory experience, we can never share identical histories. Even when we do find individuals whose comprehension appears to mirror our own, it is most likely because their perceptions reflect similarities in our experiences.

I hope you detect that this applies outside the range of mountains, but you may be wondering what all this has to do with faith. I have pursued this line of reasoning to help counter the mistaken idea that sensory experience somehow gives us unassailable knowledge and that, if we'll just rely on it alone, faith is unnecessary. In the first place, we don't necessarily even have the same sensory experience—some of us are color blind, near- or far-sighted, happen to be viewing the mountain from only one of many possible positions, and so forth. More importantly, what we believe about the world is based on discernment extending far beyond the superficial, and our visual, auditory, olfactory, and other senses are merely the gateways to the perceptions on which we build and shape those beliefs.

A vivid example of the interplay between sensory experience and belief can be seen in the Biblical account of Jesus healing a man with an atrophied hand.[30] Although belief in miracles varies widely, from outright dismissal to unquestioning acceptance, many struggle with how to interpret them. If

---

29. I see this even though none of those things are visible in the photograph. The Skillet Glacier appears prominently but the Falling Ice Glacier is obscured by part of the mountain to its left.

30. Mark 3:1–6.

you're in that third group, you may find this case fascinating because a group of individuals are motivated to attempt to eliminate Jesus, not because they thought him guilty of fraud but, precisely because they believed the miracle occurred. Yet the gospels report numerous occasions where belief in his miracles led to more worshipful responses. We can speculate about reasons for the different reactions but, as we've already seen, sensory experience is seldom the whole story.

None of this, however, is meant to demean the importance of sensory experience, which is always the basis for faith, whether we are participating in something firsthand or using our senses to apprehend it in other ways (e.g., reading or hearing about it). It is usually to our advantage to place more credence in our direct observations but this is not always the case. Science has taught us that there is plenty that we are incapable of discerning accurately via our unaided senses,[31] and we are all familiar with the need to interpret judiciously what we see on television and in movies. But sensory limitations are not our only concern. As it turns out, faith is a necessary consequence of a broad range of limitations—a key idea we will explore in a subsequent chapter.

Misconception #6: Faith is Binary

All too often, faith is referred to as if it were a binary affair—one either has it or one does not. But faith is hardly ever all-or-nothing. From the mundane to the momentous, faith exists on a graded scale. Seldom is some element of doubt absent from even the most confidently held belief. A man believes that his wife loves him with all her heart but he also knows that she loves the kids, her mother, a special friend, and chocolate (among a great many other things). What's more, she seems especially to enjoy a couple of television shows with handsome male actors. If he stops to contemplate the actual fraction of her heart devoted to him, he might discover that her devotion only merits a grade of B+. Our now sobered husband might then be led to acknowledge that his confidence in her never leaving him is really only 88 percent and not the perfect fact he had supposed before thinking about it. As we saw in the previous chapter, even an atheist may sometimes admit to the possibility, however remote, of there being a God.

Adopting a binary stance regarding faith probably appears reasonable because so many scenarios actually do have a binary persona. We accept a request for a date or not, are married or single, have or have not graduated from college, did or did not receive a particular job offer, are wanted dead

---

31. Cf. Panek, *Seeing and Believing*.

or alive, and so forth. If we aren't careful, we might begin to think this is universally applicable. It is also true that, even though one's faith in this or that is not binary, decisions that hinge on that faith may be. Consequently, a person may have all sorts of beliefs about the relative merits of accepting or declining a specific job offer but her answer must be either "yes" or "no." Her graded faith is thereby rounded off, giving it the appearance of having an all or nothing character.

A more subtle but also more insidious reason that we often apply the binary criterion to matters of faith, however, is that it simplifies things. When beliefs match or are diametrically opposed to our own in any particular area, it is relatively easy to embrace those who hold them or to keep them at arm's length. We usually feel secure when we are around individuals who share and affirm our perspectives and, although we are frequently uncomfortable around those who passionately disagree with us, we are in the habit of merely dismissing them (and their kooky ideas). Yet, when someone shares a meaningful portion of one of our beliefs but has a compelling case for another point of view with regard to other parts of it, we recognize that it places us in jeopardy. We are then faced with the prospect of having to assess our own stance while weighing the merits of theirs. Whether it is scientists faced with the prospect of a new paradigm, monotheists confronting perspectives on origins, or just John Doe contemplating social or moral dilemmas, all portend substantial effort and the frequently frightening prospect of possibly needing to revise one's convictions.

Despite our desires for simplicity, we cannot dispense with the fact that from science to religion to ethics to a husband owning up to the fact that his wife's feelings cannot be known with certainty, faith exists on a sliding scale. But it is not only our beliefs that are a matter of degree. The extent to which those beliefs influence our thinking and action, the scope of their influence on others and, most importantly, the truth content of those beliefs are all subject to a range of possibilities. Just because we have a particular level of assurance, it does not mean that our faith is justified—the hapless husband's wife might leave him tomorrow. Maybe ignorance really is bliss. Of course, if bliss is all we crave, there are a variety of pharmacological agents to which we might turn. But bliss alone is meaningless and ignorance is the enemy of any worthwhile faith. As playwright Thornton Wilder put it, "Ignorance is blindness."[32] Unfortunately, the ease of remaining ignorant is alluring. It is the primary reason that we apply the binary criterion.

There is another regrettable outcome of falling for a binary view of faith, and this is the assumption that beliefs must, inevitably, be pitted

---

32. Wilder, *Our Town*, 109.

against each other. Sometimes this appears unavoidable. It is exceedingly difficult, for example, to see how one could reconcile modern cosmology or evolutionary theory with young-earth creationism without radically modifying the typical hypotheses of one or the other. Yet, this need not be the case in every situation. Sadly, we are often prone to draw battle lines where there is really no war to be fought. As surgeon-turned-theologian Wilton Bunch has noted, many people readily accept apparently conflicting views in both science (e.g., wave-particle duality) and religion (e.g., the incarnation) but often tend to believe that it is not possible to do the same thing with regard to science and religion themselves.[33] For every Polkinghorne or Peacocke arguing for the potential compatibility of science and religion[34] there is a Dawkins or Kurtz attempting to dispute their claims.[35] But the first step toward any possibility of reconciliation is to acknowledge that every belief occurs on a continuum. That, and a generous dose of humility, can go a long way.

Finally, we ought to recognize that there is a certain freedom in the realization that faith need not be dichotomous. "Don't fence me in" is a sentiment not limited to Roy Rogers, Ella Fitzgerald, or Bing Crosby,[36] or to people facing incarceration or other physical limitations. None of us wants someone placing restrictions on his or her beliefs, but anytime we adopt the binary perspective we unduly constrain both ourselves and those around us.

Misconception #7: Faith is a Religious Phenomenon

I would like to think that I have already given this a knockout punch but it is such a pervasive belief that I prefer taking no chances. Yes, faith is a significant component of each and every religion but to allow the term to be hijacked by people (both religious and not) who want to make it synonymous with religion betrays a basic misunderstanding about what it really is. As I have attempted to point out repeatedly, faith is a factor in everything else we think and do. This has not escaped the notice of a great many people who have bothered to give it much thought. Gerald Weinberg, for example spells it out in no uncertain terms:

33. Bunch, "Two Stories."

34. Cf. Polkinghorne and Beal, *Questions of Truth*; Peacocke, *Creation and the World of Science*.

35. Cf. Dawkins, *Devil's Chaplain*; Kurtz, *Science and Religion*.

36. All of whom have performed the Robert Fletcher and Cole Porter song by that name.

> But nobody can exist without faith in something. Without faith, we could not move one foot in front of the other, not knowing whether the next piece of ground would support our weight. Moreover, we could not even stand still without faith in the continuity of the ground now beneath our feet.[37]

Josh Billings echoes this sentiment and adds a touch of humor to make his point:

> If it wasn't for faith, there would be no living in this world. We couldn't even eat hash with any safety . . .[38]

And, because science is often presented as the area par excellence where faith is frequently deemed unnecessary, if not outright disparaged, we should contemplate the early twentieth-century physicist Max Planck's insight regarding science and faith:

> Anybody who has been seriously engaged in scientific work of any kind realizes that over the entrance to the gates of the temple of science are written the words: *Ye must have faith.* It is a quality which the scientists cannot dispense with.[39]

If you are thinking about objecting that quoting supporting sources does not prove anything, I quite agree. But I do think it important to illustrate that there are viewpoints other than the one which only sees faith's conjunction with religion.[40] In the next chapter we will zero in on a characterization of faith that will, I believe, fully and finally show its ubiquity but for now we will consider several arguments that are often made when attempting to support the claim that faith is just a religious thing.

One of these is that people simply mean something different when they use the term in a religious context than when they use it elsewhere. Our use of language certainly allows this sort of thing but it also makes it easy to obscure what we are trying to say. If there is an intentional wish to muddy the linguistic waters, I have nothing else to say about this other than we are granted the

---

37. Weinberg, *General Systems Thinking*, 35.
38. Billings, *Everybody's Friend*, 46, spelling modified.
39. Planck, *Where is Science Going?* 214, italics in original. Nobel laureate Gerald Edelman claims that, "Hope and belief are as important in science as they are elsewhere; the difference is that in science they must yield to experiment" (*Bright Air, Brilliant Fire*, 208). But any valid faith must yield to evidence, even if we don't call it an "experiment."
40. Holmes Rolston suggests that, "it is quite as much an act of faith to see dinosaurs in the possibility spaces of quarks as to see dinosaurs in the possibility space of God" (*Genes, Genesis, and God*, 354). While we might argue about the accuracy of Rolston's remark, he is correct that both positions involve faith.

right of free speech (if not free understanding). But, when the objective is clarity in communication, it is important to appreciate the scope of the terms we employ. This is particularly true of a loaded word such as "faith."

A related issue is that the only sort of faith anyone is really interested in exploring is the kind that applies in the realm of religion. Clearly, there will be times when we will want to focus on faith as it applies in a certain context. However, blanket assumptions that there is but one area (e.g., religion) in which faith has appeal, that faith is a precursor to irrational behavior, or that its sole purpose is to provide motivation (and so forth) are specious and ignore the interesting and significant role it plays in all areas of life. Here's to keeping our heads out of the sand.

Finally, there is frequently a tendency to admit that, yes, faith is operative in all areas of life, but then to maintain that the degree of support is quite varied in different domains. I have no quarrel with this claim but want to make it perfectly clear that, once again, this applies to specific examples of faith and not to faith in general. Few people are surprised when they encounter fragile arguments for certain religious postures but give little thought to the fact that there is a history of this occurring even in those realms that pride themselves on avoiding such foibles. Whether it was Aristotle pontificating on the arrangement of heavenly bodies or the motion of falling objects, Franz Josef Gall promoting the merits of phrenology, Rene Descartes hypothesizing about the role of the pineal gland as the gateway to the soul, or Albert Einstein demonstrating an early blindness to the explosive potential indicated by his own famous equation ($e=mc^2$),[41] the annals of science are replete with demonstrations of misplaced faith in the possibility or impossibility of one phenomenon or another.

It is easy to simply protest that bad science does occasionally occur but that such examples should never be construed as an indictment against science in general. We must then grant that the same grace ought to be extended to other areas. Unsupported religious conviction is clearly bad religion but there is little excuse for mistaking it for a comprehensive comment on religious faith.

* * *

---

41. Isaacson, *Einstein*, 272. Einstein would later advocate for development of an atomic bomb.

In this chapter we have explored seven common misunderstandings about faith. The suggested counter-perspectives can be summarized as follows:

Setting the Record Straight About Faith

1. Faith is always based on evidence.
2. Faith pertains to the possible more than the impossible.
3. Faith and reason can and should be compatible.
4. Blindness is not a requirement for faith.
5. Sensory experience and faith must co-exist.
6. Faith is not binary.
7. Faith is more than a religious phenomenon.

Our investigation has allowed us to create vital context for better grasping the true nature and scope of faith. In the next chapter we'll consider a definition of faith that takes into account our observations so far and that sets the stage for subsequent improvements in our perspective.

— 4 —

What Faith Is: Making Faith Concrete

> For the sake of precision declaratory sentences should be formulated . . . with the words "I believe" prefixed to them.[1]
>
> —MICHAEL POLANYI

IN THIS CHAPTER I will offer a definition of faith you are not likely to find in most dictionaries but which illuminates its true nature, thereby providing insight that can enhance understanding of the term and many of its synonyms and remove some of the ambiguity frequently associated with their use. In the process, I hope to strengthen the claim that faith is ubiquitous and convince you that it is a natural and necessary part of our existence.

Faith is Ubiquitous

Through everything that has been said so far I have tried to make it clear that faith is operative in every area of life and that there are always reasons—sometimes good, sometimes not—for our faith in any particular thing. For example, consider the instances of faith depicted in Table 1.

---

1. Polanyi, *Personal Knowledge*, 299.

The	has faith that
quarterback	the receiver will catch the ball
athlete	pain means gain
student	studying will be worth it
scientist	the experiment will be productive
new employee	the job will be satisfying
guy/girl dating	this relationship could be the one
guy/girl marrying	this will enhance his/her happiness
couple planning a family	the child will be healthy; will turn out ok
religious	he/she can know God; have abundant life

Table 1: Instances of faith

In the previous chapter, we observed that religious faith is merely one kind. Here, we can extend our observation and note that it is not even the most common. For those with eyes to see, faith is everywhere. For example, Disraeli reminds us that even "Duty cannot exist without faith."[2] This is not hard to see because the very concept of duty is based on our evaluation that there will be consequences to our endeavors or lack of them, both personal and beyond. That evaluation is nothing less than faith in action.

I hope that the pervasive nature of faith has been clear to you from the beginning but, if not, perhaps by now you are beginning to have doubts about restricting it to some limited domain (i.e., such as religion). The myriad decisions we make that something is necessary or merely appropriate; the innumerable choices we make in which the outcome is unknown ahead of time—all are based on faith. Technically, this means virtually every decision is faith-based although, in some cases, we will have much higher levels of confidence than in others. Faith is active when we select a new vehicle, drive it, or take it to the mechanic. We have faith in products, peers, processes, progeny, politics, purchases, and potentially potent statements such as this. We may have strong faith in our progeny and weak faith in the political arena (or vice versa) but, either way, faith is intertwined with everything we think and do. Similar to the recursive nature of self-awareness, we even have faith about faith. Faith is inescapable.

2. Disraeli, *Tancred*, 166.

How We Know Things

As we contemplate the prevalence of faith in our lives, we are confronted, sooner or later, with the fact that what we know or think we know (or what we think we don't know) cannot really be adequately addressed until we come to grips with the issue of how we know things at all. The short answer is that our knowledge is acquired via observation.[3] However, because the form of an observation is both variable and error prone, it is useful to adopt a classification scheme whereby we identify any given example as belonging to one of three possible categories. We'll simply designate these as primary, secondary, and tertiary.

Primary observations are those in which an item or event is observed directly as the product of a personal sensory encounter with the actual entities that give rise to the experience. When we see an accident, hear the sound of rapids, smell a pizza being baked, taste goat's milk, feel the sting of hailstones, or feel that we are losing our balance, we are engaging in primary observation. This means, of course, that any observations we make of events involving ourselves—perhaps it is our own accident that we observe—are most definitely primary observations.

This can be contrasted with secondary observation in which the primary senses are still employed, but only to observe something indirectly or second-hand. We would typically classify watching an event on television, reading a book, being told about a performance, or using scientific instrumentation as secondary forms of observation. Even though there can be no secondary encounters without there also being primary ones (e.g., when we view a show on television we are directly experiencing the set itself), it is the information conveyed by our watching or listening that is of interest and that information comes to us in a secondary form.[4]

Tertiary observations take place purely internally and are the output of one's thought processes. These are the things we observe when we think about what we see or hear. Reflection and inductive or deductive inferences

---

3. Hume (*Human Understanding*) is noted for the primacy he afforded experience as the basis for understanding "matters of fact" and found it odd that philosophers had not previously paid more attention to the issue. But Harrison indicates there were individuals (including Albert the Great and Thomas Aquinas) predating Hume who advocated the importance of sensory experience (*Bible, Protestantism, and Natural Science*, 37–38). Is it possible that the role of experience (observation) is so natural and ingrained that few people before Hume thought it necessary to seek a philosopher's insight on the matter?

4. Although it is easy to suggest examples of secondary observation corresponding to vision or hearing, it is normally harder to do so for the other senses without also employing sight or sound (e.g., she *said* that it smelled like a pizza).

are examples of tertiary observation. We might note via primary observation, for instance, that a large number of the fans at a sports event we happen to attend are wearing clothes with a common color theme. The apparent conclusion would be that their attire reflects school colors but, in the absence of any other source of information to that effect, such a conclusion would be a tertiary observation. Einstein's special theory of relativity, the apostle Paul's views on marriage, calculated returns on a financial investment, and an assumption that a proposal of marriage will be accepted if it is made are all examples of conclusions based on a significant contribution from tertiary observation. There is no need to suppose that such knowledge must be shared by anyone else, even if it frequently is.

The components of our thoughts, therefore, originate in direct observation (from our biological sensory apparatus), testimony (via communication from others), and inference (as we interpret, deduce, surmise, or otherwise reach conclusions). These thoughts are the things we say we know or know about—they are the things we believe.

Another way to look at this is to notice that we are typically engaged in trying to answer two kinds of questions: "What?" and "What does it mean?" The former wants to know what something is, what happened, if it happened, and so forth, and can include any or all of the related adverbial forms involving who, when and where. Answering such questions is predominantly the role of primary and secondary observations. On the other hand, attempting to determine what something means is handled by our inferencing mechanisms and is a form of tertiary observation.

Answers to these two questions thus constitute two forms of knowledge, either or neither of which may be valid. For instance, a woman's belief that her beau went to a favored jewelry store falls in the first category and may be formed because she thought she saw him enter the store or because someone told her that he did. If she believes that he did, in fact, go, she is then faced with a host of possible meanings (whether he actually went or not):

a. He loves her
b. He is trying to seduce her
c. He owes her something
d. He went for another woman
e. He works at the store
f. He drives a delivery truck

In other words, the fact of her second belief (i.e., why he went) is contingent on the content of her first belief (i.e., that he went) and not on its veracity. Even if she is wrong about his going, she can still entertain reasons for (her presumption about) his having done so. These will be the things that she says she knows or thinks she knows.

In any event, I suspect that there may be some reluctance to use a term like "knowledge" for the results of tertiary observation. If you are one of the protestors, I applaud your insight but am obliged to warn you that your concerns should also extend to secondary and even primary observation as well. I'll delay the explanation for that claim until the next chapter but the conclusion we are rapidly approaching is that even those things we call knowledge are merely what we believe to be true. The higher our confidence in something the more likely we are to use the word, but we may be remiss in calling it knowledge. Perhaps we should call it faith. If you're not quite ready to make that leap, I can wait. Meanwhile, let's consider several important issues that arise in the context of thinking about this matter of obtaining knowledge.

In the first place, it is important to note that these components are not independent but work in both complementary and antagonistic ways. We come to certain conclusions about how the world works by incorporating what we have observed via primary or secondary means with reflection on what those observations signify (and so forth in a recursive fashion). For example, consider the scientist who, after several years of experimentation (with results obtained via primary observation) and extensive reading (secondary observation), comes to the conclusion that a particular line of research appears promising (tertiary observation), only to discover via additional experimentation (primary), comparison of results with those of colleagues (secondary), and some deep reflection (tertiary), that he was mistaken. Had he kept his thoughts to himself he might only have been thought a fool but, having proclaimed his hypothesis both loud and long, he ended up removing all doubt (in the minds of other primary or secondary observers). In any event, whether they are about experiments, investments, people, or something altogether different, our hypotheses (beliefs, if you prefer) are formed and then either supported or refuted via all three forms of observation.

Depending upon the topic and the individual involved, however, the amount of effort expended in creating beliefs can vary considerably and, consequently, so can the validity of the conclusions. Acknowledging that there is a hierarchy of validity with respect to our beliefs is not likely to shock many folks but it is important to keep in mind, particularly with regard to the three ways of acquiring knowledge. Thus, we can probably agree that knowledge obtained via primary means is typically more reliable than that obtained via secondary ones. For example, you can find any number of images of jackalopes (the fanciful cross between a jack rabbit and an antelope) on the web (i.e., via secondary observation) but you're rather unlikely to encounter one in person. In general, credibility falls off (sometimes rapidly) the further we are removed from a source (although this is not a given).[5]

Following a similar line of reasoning, we are led to the apparent conclusion that knowledge acquired via tertiary observation is even less robust than that gained through the other two forms—that anything we think we know should always be considered suspect in reverse order from how it was acquired. There is some legitimacy to this view but it tends to ignore the fact that all of our beliefs ultimately have something of a tertiary nature. I'll expand on this later but for now we can just reiterate that the further removed we are from primary observation, the harder it becomes to justify our beliefs.

For instance, some people knew for centuries that humanity was created in a day. Now, many people know that there was a slow evolutionary process instead. But each case is a matter of tertiary belief because no one from either group was there to make a primary observation. This doesn't mean that the two views are equally valid but that the tentative nature of tertiary observations should never be forgotten and that this is especially true when they cannot be validated by primary means. Many religious claims, as well as scientific observations based on the use of peripheral equipment, have this character. Although it would be a mistake to throw out religious or scientific conclusions because of this, it does provide the impetus to argue that we should put intense effort into establishing the proper framework for justifying any beliefs we deem consequential (and considering that those we don't think noteworthy just might be).

So far I have ignored one other potential mode of knowledge acquisition that should be considered if we are to be thorough—the issue of revelation. Most religious traditions not only admit the possibility of revelation but place significant emphasis on it (at least in their historic settings) and there are claims in some quarters for its continued existence.

However, whether you accept revelation wholly or with caveats, deny the prospect under any circumstances, or are merely ambivalent, it is important to consider how it might fit within the framework which we have been developing. This is easily done if there is no such thing but the term permits a range of possible interpretations. Are we talking about some disclosure from deity or merely about a level of inspiration that led to the theories of Einstein or the poetry of Tennyson?[6] If the latter, this is nothing more

---

5. You've probably played the game where a phrase is passed verbally along a human chain to see just how distorted it can become in the retelling. Usually the longer and more involved the phrase, the greater its alteration.

6. This is the same issue addressed by Dilday (*Doctrine of Biblical Authority*)

than what we have previously called tertiary observation (albeit of a superb quality). It is conceivable that many so-called revelations are actually just a notable conjunction of thoughts in a special mental context—momentous, perhaps, but not otherworldly.

If we allow the possibility that the source is outside oneself but that the information is not received via the usual senses, the closest thing we can imagine is something akin to telepathy. In such a case we might be led to extend our definition of primary observation although no one has a clue as to how this might be done (even though Descartes thought he did). However, even though most people deny that telepathy occurs, many can conceive that it might. For such people, the only reasons to deny the possibility of revelation, then, is to deny the possibility of a revealer.

At this point it probably appears that we have reached a conundrum—one either accepts revelation (or its possibility) or denies it and there is no bridging the gap between the two views. I have no desire to disregard the existence of such a divide but it is really of little consequence to our present attempts to understand the nature and role of faith. Let's assume, for example, that you are the reputed target for some revelation. In that case, you must interpret what you believe has happened to you (i.e., via tertiary observation). This whole undertaking will occur in the context of faith, perhaps beginning with an attempt to determine the likelihood that this was a genuine revelation and not just a normal tertiary process. My belief about your encounter, on the other hand, will be the product of my own reflection (tertiary observation) based on what you believe happened to you and what is communicated to me by some primary or secondary means. In short, you end up either believing that you received a revelation or that you did not and I am left to acknowledge that my response, too, is a matter of faith. The only way I can imagine to remove faith from the equation is to contend that both the message and the interpretation are revealed (to both of us) but I have no plans to pursue the idea that we are merely automatons with no reflective capabilities.

What We Know for Sure

In the previous chapter we noted that, in our rush to segregate things, we frequently take a binary view of faith—one has it or one does not. This tendency is also likely to lead us to split our mental contents into two categories

---

regarding how to understand the inspiration of Scripture. He contrasts the inspiration attributed to Shakespeare with other (presumably) less natural means. Of course different understandings of inspiration need not be mutually exclusive.

consisting of those things we know for sure and those about which there is some degree of uncertainty. But, what can one really know for certain?

At this point I would like to encourage you to make a list of seven things that you know for sure. In a couple of paragraphs you will find my response to this assignment but I am hoping that you will not peek at it until you have completed your own and followed the other instructions I will suggest below. Your list should include nothing about which you have the slightest hesitation or entertain the least doubt and, to help with some analysis that we'll want to do shortly, it should be written. Make an effort not to be redundant.

After you have completed this task (and still without looking ahead) try to characterize the things that you have recorded. Are there aspects they have in common (other than being in the same list)? Are there things you considered putting in your list but did not? Do any of those unlisted items share attributes? You might also ask yourself what your spouse or children or best friend or the President or the Queen would put in their lists and contemplate how those lists would compare with your own.

You could also speculate about the contents of my list but that won't be necessary because here it is:

Seven Things I Know:

1. I am a man.
2. I have two arms.
3. Several of my teeth have crowns.
4. I live with a woman named Carol.
5. We have five children.
6. I like peanut butter.
7. I walk to school most days.

I would love to see your list but, as that is not feasible, allow me to speculate about how it compares with mine. In the first place, I suspect that there are few, if any, exact matches. On the other hand, I would be surprised if there were not several similarities among the types of things we included. For instance, you might have expressed your certainty in being a woman, having two legs, or sporting red hair. (I could have noted in my list that I possess two legs as well but, having just listed having two arms, that seemed a tad superfluous.) In any event, if you listed one or more of your physical attributes, it may indicate that we were thinking along somewhat similar lines.

There is a good chance that you're not married to a woman named Carol, and an even better one if you're a woman yourself. On the other hand, it is entirely possible that at least one of your entries involved a relationship. Other potential similarities between our lists might include likes or dislikes (but not necessarily about food) and experiences or activities (even if you don't regularly walk to work or school).

In any event, the things in my list seem unassailable to me and, if you followed my instructions, those items in yours must appear that way to you. Here are several other items that also strike me as irrefutable and that my list might have included:

- The earth is an oblate spheroid.
- Water is soft.
- I spent two weeks last summer climbing in the Rockies.

However, there are several reasons I did not do so. For example, despite my confidence in the shape of the earth, the first statement in my "also-ran" list really only parrots what I have been taught in science classes and from other secondary means. The photographs taken from spacecraft appear to support this, as do other factors including visual evidence of curvature (apparent even to those of us who are earthbound) and a variety of reported measurements. But having never been in space myself nor made any of those measurements, I am reduced to taking the word of someone else not only for most of those claims but especially for the shape itself. I see no reason not to do so but the very fact that I do rely on someone else means I cannot honestly include the first assertion in a list of things I know for sure.

The second affirmation is also problematic and not just because water can sometimes be in the form of ice. Although as a liquid it certainly seems soft considering that we wash with it and even drink it, as I discovered a number of years ago when I jumped from a tall tree into a creek, it definitely doesn't feel that way under all circumstances. If it did, those with suicidal tendencies might have different expectations when they jump from the Golden Gate Bridge.[7]

The fundamental problem with this statement is that words like "soft" are ambiguous. In fact, many of the terms we use or the things we contemplate have a context dependent (or possibly even an inherent) ambiguity. Because there is a range of possible interpretations for statements containing such terms, even though we can (and usually do) make them with

---

7. Cf. Friend, "Jumpers."

feelings of certainty, we need to acknowledge the fundamental role played by our current perspective.

With respect to my final less-than-positive contention, I feel the most confident of all despite the possibility of some ambiguity in the word "climbing" or the fact that "last summer" will be changing from year to year. (I do not suggest that I would consider making this statement every year.) No, I am very confident that during the summer prior to the time at which I am penning these words I was climbing mountains in Colorado. The only reason I can give for not being willing to put it into the first list is that I recognize the ever-present potential for confabulation. Thus, for each of my vivid mental images of cramponing up a narrow couloir, gingerly crossing a precipitous ridge, and carefully ascending a rotten rock wall, I am aware that there is someone, somewhere, who has vivid mental images of things they have never actually done or attributes they do not really possess.[8] Technically, such concerns extend to my original list (and yours, too, for that matter), although few of us believe they apply to us.

Characteristics of Things That We "Know for Sure"

Earlier I asked you to assess the items in your list, paying particular attention to any features they have in common. This is because, when we take the time to study those things we confidently assert that we know for sure, we are apt to notice several features that characterize many of them. We are going to consider some of those shared attributes now, recognizing that some items in our lists may be characterized by only a few or even none of these attributes, but acknowledging that the search for commonalities can help us better appreciate how such items compare with all the other things we know but cannot, in good conscience, put in our lists. These attributes characterize items in my list and I anticipate that they will do so for some of those in yours as well. However, you might recheck your own list to see if it included any entries that would fall prey to concerns similar to those mentioned in the preceding paragraphs (i.e., things about which, if pressed, you should admit less than complete assurance), as such items will not necessarily possess these features. Here are some of the attributes shared by things we "know for sure":

---

8. Picture the Riddler (Jim Carrey) proclaiming from his psychiatric cell at the end of the movie *Batman Forever* (Schumacher), "I am Batman." Unfortunately, delusions aren't restricted to the cinema.

Information is Via Primary Observation

Many, if not most, of the things we say we know for sure are based on primary observation. In other words, they are "known" via direct sensory experience and are not based on information received from others (e.g., written or verbal accounts or photographs) or things we have deduced. Even observations made using scientific equipment may not be granted the same level of certainty afforded to that gained from the unaided senses.[9] Although one may have high confidence in descriptions of ultimate human origins or reports of miracles, unless one has observed them personally they should not be labeled as "things we know for sure" (even if they frequently are so classified).

Concerns Personal Attributes

In conjunction with the previous feature, we will probably discover that many of the items that might legitimately make up our lists are about us personally (discovered via primary observation). This is, of course, hardly surprising. These would be physical traits, personal experiences, relationships, tastes, possessions, and the like. As with the feature above, all seven items in my list have this property. I might say without hesitation that my office feels cold to me when I enter it but, given the vagaries of metabolic influences on temperature judgments, I cannot say so assuredly that it *is* cold unless there is a thermometer to validate my initial perception (and even then I would be trusting that the device is functioning correctly). Similarly, it would be problematic for me to say that I know for sure the sky is blue (because of what we have learned about color perception in the human brain). Yet, I can, nevertheless, say confidently that *to me* the sky appears blue (meaning by the term "blue" a color sensation that, by agreement, others also call "blue").

---

9. According to Panek, this problem confronted people attempting to use Galileo's telescope as they discovered that they must interpret what they were seeing: "how much certainty did 'sense evidence' actually offer?" (*Seeing and Believing*, 43). Or consider: "A mark appearing on a film; an electroscope discharging abnormally; that is enough to force physics to accept fantastic powers in the atom." Teilhard de Chardin, *Phenomenon of Man*, 290. Instrumentation has made possible a more accurate picture of the universe than unaided sensory experience but it is important not to discount the role of inference (and secondary sources) in obtaining such pictures.

Attributes are Verifiable by Others

Color is not the only thing on which most of us can agree. In fact, many of the things I know for sure are also things about which other people would agree with me. This is one of the reasons we can have for thinking that we're not psychotic and merely imagining things that don't exist. Not all of the items in my list have this property. For instance, there may be no one who can verify my method of transit to my office and any conclusions about my taste for peanut butter must be inferred by others based on secondary observations and a bit of inductive logic. We should also keep in mind that just because the agreement of others may characterize things in our list of certainties, it does not mean that when people (even large numbers of them) agree on something the object of their consensus is itself suitable for such a list. History is replete with scores of examples where things "known for sure" evaporated in the light of new discoveries.

Consists of Restricted Domains

The more restricted the domain, the easier it is to identify things we know for sure. For example, out of one group of students completing this assignment, several included statements such as 2 + 2 = 4. This looks safe enough except, technically, this is not true in all number bases (e.g., 2 + 2 = 11 base 3 or 10 base 4). Fine, you say, let's just limit such statements to include any or all number bases in which they are true. That, however, is my point. By restricting things suitably, we merely create tautologies which, by definition, are always true but not necessarily interesting or meaningful.[10]

One way to restrict a domain is to use hedge words such as "sometimes," "maybe," "few," and so forth. For example, I noted that, "Several of my teeth have crowns." I said it that way because I didn't remember exactly how many and was too lazy to get a mirror and count them. By hedging with "several" I was able to make a statement about which I had no doubt. Hedging can also be done without using easily identified adjectives or adverbs. I am very confident that the woman I live with is also my wife, but I haven't seen the marriage certificate in years and took the minister's word that he had the legal authority to perform the ceremony of which I remember being a part. If this seems a bit silly, I agree, but if I was faced with the task

---

10. There is no need to associate issues of certainty with a materialist position (Cf. Barr, *Modern Physics and Ancient Faith*, 201–4). Certainty by definition (i.e., knowing something to be true because it has been defined that way) is possible for materialists, spiritualists, or persons of any conceivable metaphysical bent but who cares about that level of certainty?

of ranking in order of confidence "living with" or "married to" a woman named Carol, I would keep them in that order even though there is good evidence to support both claims. Similarly, you might also have noted that I never claimed to have fathered the five children we have. Although I have no reason to doubt that they are mine, there have been no paternity tests to verify it. One thing I am relatively confident about, however, is that if my supposed wife ever reads this, my life may not be worth very much. In any event, one sometimes has to take such risks to make a point.

Involves Unremarkable Observations

One potential result of restricting domains for our statements of complete assurance is that those statements can often end up being unremarkable and relatively vacuous. This does not mean they are unimportant (although it might) but just that they are not apt to excite much interest or passion in any one to whom we make them. For instance, my statement "I am a man," although of great significance to me and something about which I feel perfectly confident, is not likely to leave you exclaiming, "Wow, Elma, did you know that the person who wrote this book is a man!" Other possibilities (e.g., "This list contains seven items") may meet the required criteria but be truly trivial. Look at your own list and see how many of the entries have this characteristic.

What We Know for Sure (Redux)

As we have begun to see, there may be things we claim to know for sure whose justification can be problematic. When perusing actual lists of individuals asked to complete this task, I have discovered that this conclusion is inescapable. Furthermore, even though you may have been careful to exclude such items from your list, you have probably encountered a number of persons whose claims of assurance in one thing or another you were prone to doubt.

For example, a student who anonymously completed this exercise listed the following for the first two items: (1) God is my Savior. (2) Nothing is certain. Although there are a variety of things we might say about these, the most obvious is that, if the student knows the second for sure, he/she cannot know the first with the same degree of confidence. Hopefully, there are few conflicts between things we might mention, but let me ask you to explore this idea of certainty further by ordering the following seven items on the basis of your confidence in them:

- Everything happens for a reason.
- My parents love me.
- A man has been to the moon.
- We will all die.
- Computers know nothing more than 1's and 0's.
- Things never turn out exactly the way you expect them to turn out.
- The area of a circle or square can be known definitely.

Each of these, it turns out, appeared somewhere on the lists of people who were asked to complete the task you performed earlier. In other words, someone is willing to say they know at least one of these things for sure. Presumably that means, for the person making such a declaration, that there is no possibility whatsoever that some things might just happen (i.e., with no teleological overtones), that their parents are merely keeping up appearances, that the moon landing was staged, that someone might not die, that Deep Blue (IBM's chess playing computer) knows more than binary code, that sometimes someone might actually predict precisely how things will turn out, and that, if measurement is involved, the area of any actual figure can only be approximated.

It seems safe to say that one thing which can be claimed with certainty is that there are people who are willing to make all kinds of claims *they* believe are absolutely, unequivocally true. My contention is that most such statements should actually be considered less than certain. If you were able to rank the items above in any way at all, you have supported that point.

In any case, the possibility of being wrong doesn't keep some individuals from proclaiming (and perhaps even feeling) complete certainty. Although many of the things that we consider truly meaningful are actually held with varying degrees of assurance (and cannot, therefore, honestly be listed as things we know for sure), it is usually convenient to pretend that they are indisputable facts. Doing so is comfortable, expedites decisions, may aid communication, and helps us fit in certain groups. Thus, even though there are many things we cannot legitimately include in our lists of certainties, we'll still live as though we know them for sure. I have no basic quarrel with this posture as long as it is recognized for what it is. When it is not, things sometimes get sticky and we can find ourselves attempting to defend a vulnerable position.

If you have followed all this closely, I suspect you may have noticed that there are at best a small number of things we actually know for sure compared to the total number of things we say we know. We know plenty, but the bulk of that knowledge is contingent, graded, hierarchical, and open to the possibility (even if only slightly) of refutation. Consequently, when we say we know something, there is an excellent chance that we are really making a statement of faith. There may really be some things we know. There are a great many others that we only believe.[11]

Following the Odds

To this point I have mentioned faith repeatedly without ever defining it explicitly. I have avoided doing so because the things we have been discussing form the framework on which to build an understanding of faith that is as accurate and complete as possible. Consequently, I have relied on your intuitive understanding of the term and the context in which you saw it here to create an implicit characterization. I am now ready to state it explicitly. Based on all we have discovered so far,

> *Faith is the probability assigned by an individual to the existence of something (including the nature of its attributes) or to the occurrence of an event (past, present, or future), or to the reason for the existence or occurrence of something.*

There you have it. I hope that doesn't seem anticlimactic but once this is clearly grasped it becomes particularly apparent that faith is operative in everything we do.[12] That was the claim made at the beginning and which the intervening chapters have been written to support. In light of this concept of faith, the misconceptions considered earlier become quite obvious. For example, because faith is a form of probabilistic understanding, it will always have some evidential component[13] (even if that evidence is poor or fabricated). In fact, the majority of things toward which we have faith will occupy the realm of the possible (not the impossible), if for no other reason than faith cannot be separated from a majority of the things we say we know. Furthermore, when we acknowledge the graded nature of faith, we immediately see how reason and evidence can move our confidence in something up or down the scale. We also become aware that blind faith is

---

11. Cf. Locke (*Human Understanding*, book IV, chapter XV, #2): ". . . most of the propositions we think, reason, discourse—nay, act upon, are such as we cannot have undoubted knowledge of their truth . . ."

12. Contra Adams (*God's Debris*), I have no intention of making a god out of probability.

13. Cf. Bartholomew ("God and Probability" in Watts, *Creation*, 139), who describes two types of probability, one of which "expresses a degree of belief and is a measure of the strength of the evidence for some proposition."

of our own choosing and can be surmounted by employing the senses (both primary and secondary) and reasoning faculties (tertiary) with which we have been equipped. Consequently, because faith exists on a continuum, it usually makes little sense to make claims that someone either does or does not have faith. When such statements are valid, they are merely ways of saying that the probabilities are judged to be one or zero.[14] And it should be clear by now that religion cannot steal the term for its exclusive use.[15]

In a nutshell, faith is probabilistic observation (in the sense in which "observation" was used earlier in this chapter). If probability conjures up unpleasant mathematical memories, replace it with "likelihood" or "percent" or anything you please as long as the term you choose denotes the fact that faith in something is simply a reflection of the extent to which we accept it as true. Faith is really just an answer to the question, "What are the odds?"

Let's make this concrete with a simple example where we attempt to answer the question, "What is the ideal weight for a typical marathon runner?" I've phrased the question this way because, without identifying a particular individual or race, there are a variety of factors—sex, percent body fat, muscle composition, VO_2 max, race course topography, and so forth—that prevent us from identifying any specific weight with complete confidence. Consequently, we must make a best guess but do so knowing there is some likelihood that there is a better one. The chance we assign to our best guess is its probability. We would also expect the weights closest to our best guess to have a reasonable possibility of being optimal if ours is not, and the odds we give those weights of being ideal are just their probabilities. The further we get from our original guess, the less likely we are to believe that the associated weight represents a realistic answer to the question. Indeed, all possible human weights will have some probability, although most will be extremely small. Figure 1 shows a fanciful example of how we might graph our assignment of probabilities for this task. The higher probabilities hover around the best guess with the remainder tailing off rapidly in both directions. This graph reflects in a quantitative way our intuitive feelings about one of the ingredients in marathoner success.

---

14. Cf. Locke, *Human Understanding*, book IV, chapter XV, #2 on the extremes of our confidence of probability assessments from "certainty" to "impossibility."

15. Morowitz refers to five source of knowledge, of which faith is one (*Emergence of Everything*, 189–91). But faith is not a source of knowledge. It is, instead, an assessment of what one would like to call knowledge and as such is inextricably intertwined with each of the other four items.

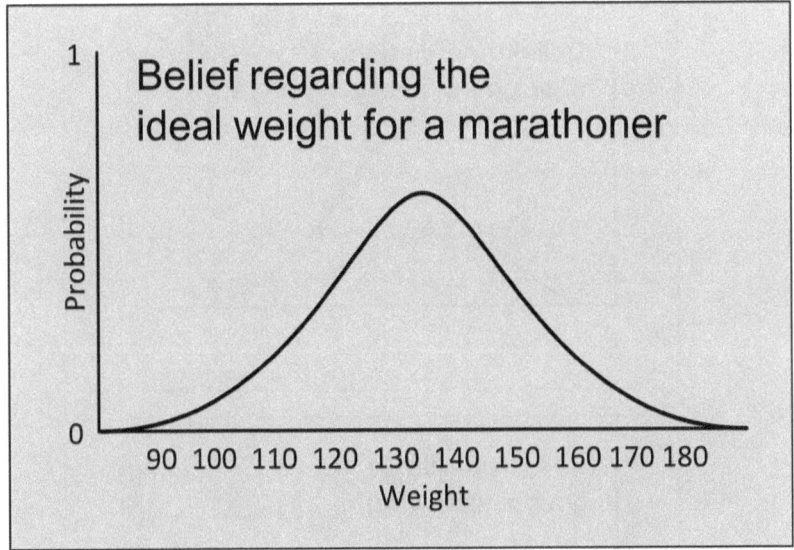

Figure 1. An example involving a hypothetical belief regarding the ideal weight for a marathoner

It is also important to note that, in the process of assessing the probabilities in this task, we may very well be relying on another set of probabilities pertaining to the relative significance of each factor mentioned above. For example, if we feel course topography is the major contributor to a determination of ideal weight and believe that most marathon courses are relatively flat, we may adjust our answer toward a higher or lower weight that reflects those biases.

Although you may profess little experience with marathons, the example above is only one of an almost unlimited number that could be put forth in a visible fashion to demonstrate the probabilistic content of faith. Figure 2, for instance, simply re-labels the axes to move us into other—for most people—more relevant domains.

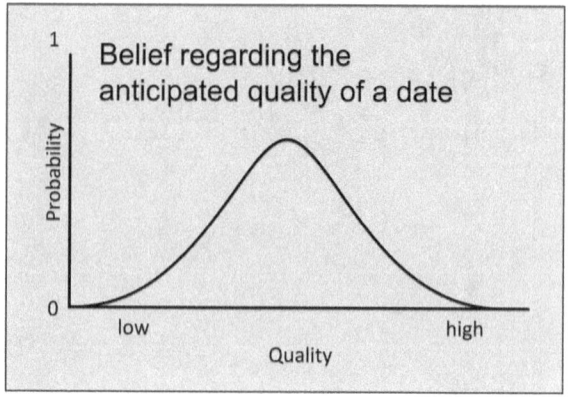

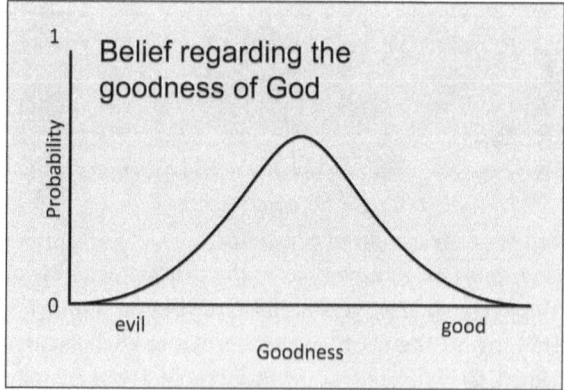

Figure 2. Modified versions of Figure 1 to illustrate the common, probabilistic content of faith across multiple (possibly unrelated) domains

The locations, heights, and even the shapes of the curves in Figures 1 and 2 can and should be modified to reflect the unique beliefs of each individual with respect to the indicated areas, but that won't change the basic fact that beliefs are probabilities. Furthermore, our language is laden with words that implicitly show how thoroughly the probabilistic nature of faith permeates our conversation and thought: She *probably* loves me. *Maybe* that was a miracle. That *could be* a good investment *assuming* the market remains stable. It was *possibly* the worst performance I have ever seen. *Perhaps* he intended something else.[16]

---

16. Studies by Simpson ("Specific Meanings") and Hakel ("How Often is Often?") indicate how many of our words or phrases can be viewed in probabilistic terms. "Fuzzy logic" attempts to exploit this basic idea for facilitating intelligent reasoning with computers (cf. Zadeh, "Fuzzy Logic"). For a general introduction to fuzzy logic (referencing the Simpson and Hakel studies), see McNeil and Freiberger, *Fuzzy Logic*.

Aside from wagers, however, we seldom actually assign a numeric rating to our beliefs. But every belief is a type of wager. Pascal's is the best known—". . . you must wager. It is not optional. . . . Let us weigh the gain and the loss in wagering that God is"[17]—but, although arguably the most important, probably not the most common. We are constantly wagering about how someone will respond to us, if we'll get seated quickly at a restaurant, whether we can find an appropriate gift within our budget at a particular department store, or if eating that last pastry might put the finishing touches on some nearly clogged artery.

Observation about the integration of probability into our decision processes is not a new idea. Over two thousand years ago, Cicero noted that "many things are *probable*, and . . . though these are not demonstrably true they guide the life of the wise man, because they are so significant and clear-cut."[18] The existentialist philosopher Jean-Paul Sartre simply noted, "When we want something, we always have to reckon with probabilities."[19] Whether the things that guide us are "clear-cut" (as Cicero maintains) or not, faith is our rating of just how clear-cut those things are.[20] Locke claimed, "Probability is likeness to be true, the very notation of the word signifying such a proposition, for which there be arguments or proofs to make it pass, or be received for true. The entertainment the mind gives this sort of propositions is called *belief, assent,* or *opinion . . .*"[21]

17. Pascal, *Pensées*, #233. We'll examine this closely in a later chapter.

18. Cicero, *Nature of the Gods* (Walsh translation, book 1 # 12, 7, emphasis in translated version). Rackham translates the first phrase as, "many sensations are *probable*" (Cicero, *De Natura Deorum*, 15).

19. Sartre, *Existentialism*, 29.

20. Scientist and author Michael Shermer notes, "In science, claims are not true or false, right or wrong in any absolute sense. Instead, we accumulate evidence and assign a probability to a claim . . . The same way of thinking has application to morals and ethics" (*Science of Good & Evil*, 167). Even IBM's Watson, designed to excel at the game *Jeopardy*, relies on a probabilistic approach for selecting its response (Cf. Feldman, "Must See TV").

21. Locke, *Human Understanding*, book IV, chapter XV, #3 (emphasis in original). Locke notes a relation between probability and faith and also points out (ibid., 4) that, "The grounds of probability are two: conformity with our own experience, or the testimony of others' experience." This sounds like what we are calling primary and secondary observation; elsewhere (ibid., book II, chapter I, #4) he describes the tertiary means. So far, so good. Unfortunately, despite the apparent overlap in our beliefs to this point, it appears that I must part company with Locke, who goes on to claim (ibid., book IV, chapter XVIII, #2) that faith, "is the assent to any proposition, not thus made out by the deductions of reason, but upon the credit of the proposer, as coming from God, in some extraordinary way of communication." Clearly this is a much reduced view of faith from that described in this book and reflects some of the misconceptions discussed in the previous chapter.

It can be difficult but also revealing and sometimes even a bit disconcerting to actually attempt to characterize one's beliefs in terms of probability. Let's try it. I will suggest several areas of belief that I imagine will be pertinent to most readers. Your task is simply to assign a probability value to each. If this seems familiar it is because we did an implicit version in the previous section. Here, however, you are not being asked to provide a relative rating but to perform an absolute assessment. We'll let an assignment of 1.0 mean that you hold the belief with complete certainty and feel equally confident that you can justify doing so (i.e., it could have been one of the items in the list you compiled earlier). You'll use a value of 0.0 to indicate no faith whatsoever in the proposition. I expect that most responses will fall somewhere between these extremes but that's up to you. Here are the statements:

1. This is a good time to invest in the stock market.
2. _____ loves you. (Try it for several different people.)
3. Human consciousness is far different from that of any other animal.
4. The long-term effects of genetic engineering will be positive.
5. Green Bay will win the Super Bowl next year.
6. Humans have free will.
7. You will live to be 90.
8. God exists.
9. God is always good.

Once you have completed assigning values to each of the listed items, you might also try adding a few entries of your own. This will be most illuminating if you select beliefs in areas about which you are conflicted. Those shouldn't be too hard to identify for most folks but I'll warn you that it could be painful.

I've attempted, in this short list, to include a range of item types because it will help provide several insights into the process of actually thinking about beliefs in terms of probabilities. In the first place, it is apparent that such items are not of equal significance. A higher probability of your living to ninety (assuming you aren't already there) is likely to be of more import (to you) than the same probability for some athletic tour de force. Similarly, a .333 chance of a baseball player getting a hit (i.e., his batting average) is viewed quite positively whereas the same odds for an investment opportunity would be considered appalling. Of course, the importance of some of the things we believe are contingent on special circumstances. For example, unless one actually has money to invest, an appraisal of the chances of doing so is merely academic. Other issues, however, such as computing odds in Pascal's wager, can directly affect one's lifestyle and have the potential for a longer lasting impact. And then there are those situations in which some beliefs are largely determined by others. Richard Dawkins, who (as we noted earlier) gave God a .167 chance of existing,[22] is not likely to spend too much time evaluating God's goodness. Finally, it should be fairly obvious that the probabilities you assign to some of these will depend upon when you do so. If you are reading this book during a personal economic crisis, you may be less likely to believe it wise to invest than you would in a time of prosperity. As it turns out, however, accumulating (or forgetting) evidence means that the probabilities associated with much of what we believe will fluctuate with time.

If you spent any time at all assigning probabilities to the items in the list, I suspect that you may have found it difficult to assign a value to some of the statements—at least a value with which you were comfortable. There are several possible reasons for that. It may have simply been that you don't feel as though you know enough about something (such as genetic engineering) to make a valid assessment. This will not necessarily prevent you from having a belief about it but certainly points to its tentative nature. This is important because there can actually be a great many things about which we have strong beliefs but are really quite ignorant. When we reassess such beliefs in the light of things about which we are more confident, we may be faced with the need to modify our ratings. This kind of reassessment is problematic, however, because a collection of beliefs that are weakly supported by trustworthy evidence can, nevertheless, reinforce one another in an almost impenetrable web of illusion.

It may seem strange, but another reason that assigning probabilities to our beliefs can be difficult is because we may know too much.[23] I don't mean to suggest that too much knowledge is a bad thing, but sometimes we have studied a subject so thoroughly that it is tricky to know where we truly stand. For example, many people will parrot a learned response about free will but those individuals who are aware of the intricacies of the subject—including implications from physics, neuroscience, and theology—may be less quick to quantify their belief about it.

Furthermore, it is easy to see that, based on our definition, faith in any particular thing is always accompanied by a complementary faith in the

22. Dawkins, *God Delusion*, 50–51. On his own 1–7 scale Dawkins classified himself as a 6. With 1 representing a probability of 1.0 that God exists and 7 representing a probability of 0.0 that he does not, Dawkins's rating corresponds to a probability of .833 that God does not exist or .167 that he does (i.e., there are six intervals).

23. Berry, *Life is a Miracle*, 149.

opposite of that thing. Thus, the weather forecaster who maintains that the chance of rain tomorrow is 75 percent is implicitly acknowledging that the chance of it not raining is 25 percent. Furthermore, as one probability goes up, the other goes down. This may not be too worrisome when it comes to precipitation but if a statement such as, "I think she loves me!" is translated into a quantitative assessment with a probability of .75 that she does so, it also means there is a one in four chance that she does not. That may be something we don't want to know.

Ultimately, then, we can only assess any of our beliefs accurately if we are willing to allocate sufficient effort. Reluctance to do so can simply be a matter of time and resources but it can also be the result of being in turmoil about those beliefs. In other words, it may be more comfortable to keep before us what we might term "wishful probabilities" instead of those that reflect our real beliefs. For example, it can sometimes be difficult to face up to the likelihood that a relationship, personal attribute, religious principle, or even a scientific hypothesis might not have the characteristics we desire to say it does. Occasionally this is a matter of not having made an effort to determine where one is positioned on the relevant belief scale but it can also occur when individuals pretend to believe something to an extent that transcends their genuine beliefs.

Regardless, people in either group can use their ignorance to hide the fact that their verbalized beliefs do not match their actual ones. Failure to attempt to get a handle on real probabilities permits the maintenance of appearances (both for others and oneself) but inhibits forward progress. It's a bit like a person who suspects they are overweight but is hesitant to verify or refute it via an actual measurement. Finding a scale takes some effort but even greater exertion looms over the message it might reveal. Nobody said it would be easy, but who really wants to live their lives on bad or unknown premises if there is the possibility for something better?

We'll have much more to say about this later but, as we conclude this chapter, I want to reiterate that the exercises it contained were intended to illustrate the probabilistic nature of faith. Whether you pondered those items with pen and paper or just made a few quick mental notes, I suspect you couldn't help noting that your beliefs exist on a continuum. What you may not have noticed, unless you took some pains to be thorough with your responses, is that when we attempt to uncover the underlying quantifying description of our beliefs, it becomes much harder to hide the real extent of our faith (in whatever is currently on the table). The nebulous character of the usual qualitative approach may have a certain charm but this view of faith can put us in closer contact with what we really believe. We just need to be prepared to handle it.

——— CHAPTER 5 ———

Faith and Brains

> And still they gazed, and still the wonder grew,
> That one small head could carry all he knew.[1]
> —OLIVER GOLDSMITH

IN THE PREVIOUS CHAPTER I suggested that the most productive way to appreciate the nature and role of faith is to view it as a probabilistic component of (or response to) observation, where observation takes on all of the forms by which we understand ourselves to "know" things (including the conclusions drawn from thinking and reasoning). Because faith's subject matter is everything (and not even predominantly religious), it seems that anyone interested in successful living would desire to maximize the probabilities. That will be the focus of a later chapter but I proposed in the previous one that considering the quantitative component of any belief (i.e., actually assigning a numerical value to the probability of its being true) can be instructive if one is willing to assume the risks.

Facing the threat of being moved from our comfort zones is primarily a matter of courage. However, there is also the danger that the effort could tax our means, raising the specter of how one avoids falling into an epistemological abyss, forever rehashing all the philosophical dilemmas of knowing. Yet we need not worry too much about that. Although few of us will ever spend much time consciously assigning probabilities to our knowledge content, those probabilities are being assigned automatically for us all

---

1. Goldsmith, *Deserted Village*.

the time. That's what brains do. Far from being some esoteric commodity reserved for a few weak-minded zealots, far from being a necessary evil, faith is the natural product of human brains. As a result, the low-level neural bases for all beliefs are the same. Content and location (brain region) vary, but the same basic type of cognitive hardware supports beliefs about relationships, science, religion, and anything else you care (or don't care) to mention. In the following sections I will try to make it clear why this is so and what it means.

A Memory in the Making

All of the things we know or think we know are the result of some particular configuration and operation of our cerebral machinery, much of it involving our memory storage and retrieval mechanisms. I could not have written that sentence without recalling what I have learned about how brains work. Neither, however, could I have done so unless somewhere in my brain was a record of the alphabet, a collection of grammatical conventions and spelling rules, and instructions about how to utilize a computer and some specific software to transcribe my thoughts. It doesn't end there, however. Both my brain and yours, at this very moment, are relying on memories that underlie our ability to read and even to process the visual scene before us. As you might suspect, we are only conscious of some of the memories at work while others do their job largely unnoticed.

The very concept of memory has been sliced and diced in any number of ways in an effort to describe what we know about it from both physiological and psychological perspectives. The literature is replete with characterizing adjectives such as working, short-term, long-term, explicit, implicit, semantic, declarative, and procedural.[2] We speak of genetic memory to express our understanding that brains come prewired for certain tasks and thereby to refute any tendency to take Locke's tabula rasa views too far.[3] Notable features of memory are sometimes given special names—for example, "flashbulb memory" to describe those enduring recollections of our experiences occurring in conjunction with some super-memorable event such as an assassination.[4]

---

2. Cf. Baddeley, "Working Memory"; Desimone, "Physiology of Memory"; Roediger and Goff, "Memory" in Bechtel and Graham, *Companion to Cognitive Science*. For an extensive list of memory modalities including those both sensory and proprioceptive, see Baron, *Cerebral Computer*.

3. Locke, *Human Understanding*.

4. Cf. Pillemer, "Flashbulb Memory."

Although most of these distinctions need not concern us here, they suggest that a vast amount of attention has been devoted to understanding memory and serve to remind us of its critical importance. Memories are our history but, strip them away, and we're history. The devastating effects of strokes, dementia, and other brain trauma illustrate this all too well.[5] In short, at a most fundamental level, we are our memories.

Without memory, then, there is no belief in anything. Indeed, without memory, there is no belief. But beliefs are more than the mere products of memories—beliefs are memories. Or, because of the probabilistic operation of brains, we might say memories are beliefs.

The Push and Pull of Things Observed

In this section we consider several reasons for asserting that the brain is a probability machine.[6]

The Nature of Perception

Modern neuroscientists have pointed to the less than complete reliability of perception and argued for the probabilistic operation of low level functions in sensory circuitry such as that found in the vision system.[7] Various cerebral mechanisms that have been configured through a combination of genetic programming and environmental influence operate below the level of consciousness to facilitate processes such as the detection of edges or color perception which can vary with time and circumstances both between individuals and within the same brain.

Many years ago the apostle Paul wrote, "we walk by faith, not by sight."[8] The theological overtones in this statement are outside the scope of our current emphasis but there is a certain neurological legitimacy in his observation. In fact, it would be completely accurate scientifically to say "we walk by faith in sight."[9]

---

5. The computer may not be a perfect metaphor for the brain (i.e., its processing schemes and memory organization are quite different) but its dependence on memory is comparable. A computer without memory is worthless and so are we.

6. This is not quite as harsh a sounding categorization for brains as Minsky's "meat machines" which, despite the pejorative overtones, reflects the fact that there are things our brains just do for us automatically.

7. Cf. Purves and Lotto, *Why We See*.

8. 2 Cor 5:7, Revised Standard Version.

9. Had he known more about the brain, Aquinas might have decided to rethink his

In his book *Mountain Light*,[10] renowned mountaineer and landscape photographer Galen Rowell describes one of his images in which the grass in an alpine setting had a captivating color he seemed unable to duplicate in subsequent shots. When it eventually dawned on him to try viewing the grass alone (that is, by excluding its surroundings), he discovered that its color was no different than that in other photos. It was the visual context that rendered it exceptionally vivid and not the grass itself. I'll have more to say about context in a moment but, for now, Rowell's insight about the relative nature of color illustrates how our brains can make decisions automatically and below our level of conscious awareness about something as basic as "green."[11]

This phenomenon is not restricted to vision. A particularly interesting example comes from the realm of language learning. Non-native English speakers sometimes struggle with certain English language sounds despite having lived for years in a country where English is the primary language. Their children, however, if raised in the presence of normal English speakers, have no trouble producing the correct sounds. Speech scientist Patricia Kuhl has hypothesized a "perceptual magnet effect"[12] where a critical period during which the brain is wired to process speech sounds is followed by a stage in which sounds that were not part of the original learning environment will be heard as sounds from the speaker's native language. In other words, the brain develops an attractor for speech sound X during the early months of life but, afterwards, when exposed to speech sound Y will "hear" X instead (if sounds X and Y have some requisite similarity with respect to the formation of neural connections). The mistake in hearing, then, translates to an error in speech. Most native English speakers have observed the tendency to replace "l" with "r" by many Asians for whom English is a (late-learned) second language, but Kuhl's theory would apply to all of us.

The take-home point is that it is not just our high-level reasoning that has a probabilistic character. Even primary observation—that part of our interaction we are most inclined to believe is totally objective and hence safe from faith—is probabilistic.

---

view that, "neither faith nor opinion can be of things seen either by the senses or by the intellect" (Aquinas, *Summa Theologica*, volume II, part II, "Of Faith," article 4). What he calls "opinion" we might label "low-probability faith."

10. Rowell, *Mountain Light*, 125–6.

11. Regarding Michael Faraday's success in debunking a supposed instance of paranormal activity, mathematician Leonard Mlodinow notes that, "Human perception, Faraday recognized, is not a direct consequence of reality but rather an act of imagination" (*Drunkard's Walk*, 170).

12. Kuhl, "Learning and Representation," 813

Neuron Behavior

When considering the role of built-in, probability-oriented mechanisms, we could proceed all the way to the quantum level but that will not be necessary. The influence of quantum events on mental processes is an open question[13] but, whether it occurs or not, we need go no further than the neurons themselves to appreciate the integrated, low-level, sub-conscious role of probability in brains.

All of the mental operations we label as recall and thinking are an elaborate signaling process between neurons. This process is primarily electrochemical in nature and is based on voltage changes within the neurons. Whether any particular neuron at any given time plays a role in signal transmission or reception is determined by a host of molecular factors both within and outside the cell.[14] Alter some of those factors and the neuronal response may also change.

Consequently, although similar sets of neuronal constraints (inputs, cellular configuration, extracellular content) typically produce comparable responses in a neuron, those responses can vary. In the extreme, one could imagine the presence or absence of a single neurotransmitter molecule making the difference in whether a neuron fires or not, although it is seldom quite that dramatic. The point is that, under a given set of circumstances, there is only a probability that the neuron will or will not fire. Although many of these borderline effects will potentially cancel one another at both the single neuron and neural population levels, the concurrence of a few such events could result in changes at higher levels (perhaps even impacting conscious processing), meaning those are also probabilistic.

The bottom line is that faith is not just a matter of our consciously weighing alternatives to decide what we believe. If probability is part and parcel of what we are at the most fundamental physiological levels, then we have no choice but to acknowledge that the things we know—really, the things we believe—are built on that probabilistic foundation.

Contextual Influences

As neurons fire, symbols are created, stored, and retrieved. On this foundation, thoughts emerge, interact, and elicit other thoughts. These thoughts are

---

13. Cf. Penrose, *Emperor's New Mind*; Stapp, "Quantum Mechanical Theories of Consciousness," in Velmans and Schneider, *Companion to Consciousness*; Walker, *Physics of Consciousness*.

14. Cf. Robbins, "Pharmacology of Thought and Emotion"; and Bliss, "Physiological Basis of Memory," in Rose, *Brains to Consciousness*.

what we believe about the world, including what we think we know about ourselves. The brain states corresponding to an ensemble of such thoughts occur in a rich context of neural activity where, according to psychologist Thomas McNamara, "expectations and . . . perceptions . . . are, in fact, just unconscious probability judgments based on past learning and current, incomplete sensory data."[15] Actually, cellular context gives rise to a higher level context of both unconscious and conscious influences that constitute the cognitive background against which our beliefs are formed.

Philosopher of science Karl Popper has pointed out that even our most basic observations about the world are made within the framework of some prior hypotheses about what our brains expect to find.[16] When I was a college student, I spent a mini-term one January at a marine science facility located on Dauphin Island, situated adjacent to Mobile Bay and the Gulf of Mexico. One blustery day on the north side of the island, one of the strangest birds I had ever seen caught and held my attention as it flew along the shoreline. As I struggled to identify it, it finally dawned on me that it was nothing more than a common seagull but that it was flying backwards. Its feat, however, was not due to some extraordinary acrobatic skill but to the fact that it was flying into a headwind sufficiently strong to push it in a direction opposite to its intended course. Based on what I took to be the simple truth that birds fly headfirst and, having never previously seen a bird execute such a maneuver, I had incorrectly assumed that its tail was its head and vice versa—no wonder it looked strange!

But this is just Popper's point. What I expected to see (i.e., a bird flying headfirst) exerted considerable pressure on what I actually believed I was seeing (i.e., a strange bird with a recessed head and bulbous tail). The fact that I recovered from that erroneous belief is little consolation with respect to those about which I am completely unaware. The simple fact is that all of our experiences, thoughts, and beliefs are filtered through and manipulated by our memories of previous experiences, thoughts, and beliefs. Because there is no requirement that all of those memories reach consciousness, we are seldom aware of the extent and power of our histories with respect to the pressure they exert on our perceptions and convictions.[17]

This push and pull of our cerebral milieu launches us on some cognitive trajectory which is subject to perturbation due to the presence, absence, or changing influence of a plethora of factors. It is, therefore, relatively easy to

---

15. McNamara, *Evolution, Culture, and Consciousness*, 130.

16. Popper, "Philosophy of Science: A Personal Report," in Mace, *British Philosophy*, 155–87.

17. Cf. McNamara, *Evolution, Culture, and Consciousness*, 78.

imagine how any particular belief could be shifted toward greater or lesser degrees of confidence depending upon the actual factors working at any moment. Figure 1 illustrates this concept and suggests a simple physics analogy.

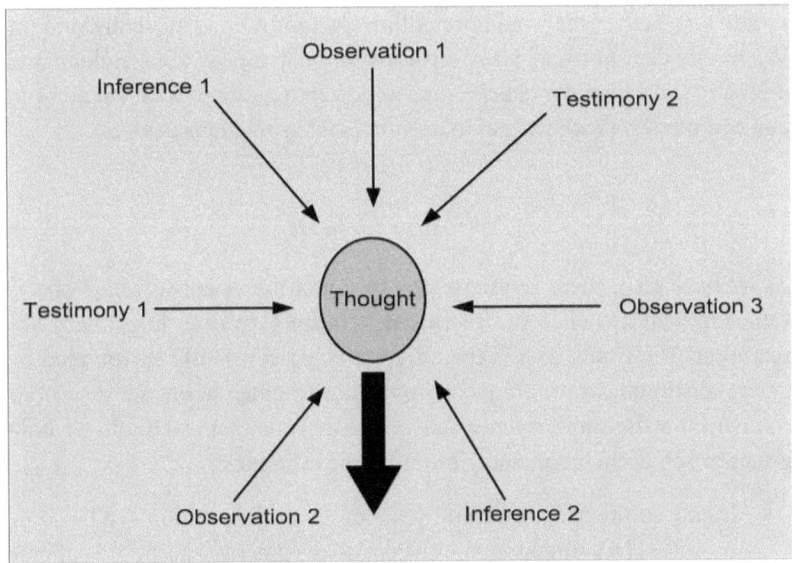

Figure 1: A force field model of thought trajectories. The trajectories of our thoughts are determined by a plethora of factors including direct observation, testimony, and inference.

In the diagram, each of the thin arrows corresponds to some factor that plays a role in determining the content of a particular thought which is designated by the shaded shape in the center. Depending upon the characteristics of each factor (itself likewise influenced by a variety of factors), the content of the thought under consideration can develop in any number of ways. The heavy dark arrow suggests one possible trajectory for that thought or belief. Primary, secondary, and tertiary observations (as described in the preceding chapter), then, can be viewed as mental forces which, like a collection of magnetic forces (with specific magnitudes and orientations) influencing the motion of a metal ball, exert an analogous pressure on the character of our belief. This mental tug-of-war, of course, has its origin in the excitatory and inhibitory influences at the level of individual neurons.[18]

---

18. The IAC neural network model provides an excellent illustration (McClelland and Rumelhart, *Explorations*).

In short, no belief exists in a contextual vacuum. As a result, context impacts content with the potential for a cognitive butterfly effect[19] leading to mental states that can best be understood probabilistically.[20] This is easily recognized in the myriad shifts in our beliefs about mostly trivial issues (e.g., whether she would prefer flowers, candy, or a frugal husband for Valentine's day) but may also help explain what appear to be sudden and dramatic shifts in major beliefs (such as perceptions about what would be in our best interests with respect to occupational or religious choices).

Attention Deficits

As we have seen, there is little reason to doubt the extent to which observation influences belief. But there can be huge gaps even in primary observation. The simple fact is that, despite being constantly surrounded by potential stimuli, we are frequently oblivious to much of our sensory environment. Furthermore, even when we are aware of some stimuli, we only get a portion of their content. Here are some examples:

- Innate auditory aptitude and years of musical education can be compromised by something as transient as an earache, leading the sufferer to misjudge the quality of a symphony.
- Most people are familiar with the so-called "cocktail party effect" where, even in the presence of multiple simultaneous conversations, it is possible to attend to one of them but impossible to concentrate on them all. Think of all the gossip that is missed!
- Look at any relatively complex image or visual scene, then turn away and try to identify all of the components in it. Better, yet, try to draw it as accurately as possible. When done, look back at the original and notice all the elements you omitted. Those are the gaps in your primary observation.[21]
- Without my reading glasses, I can process half the amount of text on this page in a given amount of time that I can with them.

Our beliefs about the quality of a piece of music, potential for interesting conversation at a party, presence or absence of an item in some environment,

---

19. Cf. Lorentz, "Deterministic Nonperiodic Flow."
20. Cf. McNamara, *Evolution, Culture, and Consciousness*.
21. As Chabris and Simon have document (*Invisible Gorilla*), even the most extraordinary things can be missed due to a phenomenon termed "inattentional blindness."

or significance of some literary work are not only influenced by our backgrounds but by a basic inability to collect all of the available information in our surroundings.[22] Certainly this problem is not restricted to sight or hearing but extends to all of our sensory systems.[23] These input constraints thus contribute in unpredictable ways to the push and pull on cognitive states, further reflecting the probabilistic nature of our mental processes.

The Adaptive Nature of Faith

Driving by a local drugstore one Sunday morning, I noticed its digital signboard proclaiming that it was 7:12 a.m. on January twenty-third and that the temperature was 505 degrees Fahrenheit. That seemed a bit warm for January, even in the South. However, it was not January. In fact, the rest of the country was preparing to celebrate Independence Day on Monday. Furthermore, it was a few minutes after 9:00 a.m. and I was about to pick up my last load of passengers before heading for Sunday services. But how did I know all this? Why were those signboard data unbelievable to me (and probably anyone else who might have noticed them)?

Fortunately, my belief system, while not flawless, gave the impression of working fairly well that day. I was not inclined to go running off to buy an asbestos suit, pine away wondering how I lost six months of my life if it really was January, or go back home for an extra hour of sleep. In the presence of a faulty belief system, however, I might have exhibited any of those responses or a variety of other behaviors that were at best merely harmless (perhaps stepping up the crusade against global warming) but which, in the worst case, could lead to any number of disastrous consequences (e.g., crashing the bus so as to invite a quick death, thereby escaping the anticipated trauma of being roasted alive).

Faulty beliefs, evidently, may lead one to think or attempt almost anything and sometimes that happens. By and large, however, our belief systems do a credible job of keeping us on track. The reason is quite simple. If faith is the probabilistic product of brains as I have been suggesting, and if brains are supposed to reflect a positive adaptation to our environment, then faith itself must be adaptive.[24] Some would probably prefer to admit only that

---

22. Cf. Churchland et al., "A Critique of Pure Vision," in Koch and Davis, *Large Scale Neuronal Theories*, 23–60.

23. Cf. McNamara, *Evolution, Culture, and Consciousness*, 130.

24. This operational (or functional) depiction of faith complements the descriptive definition given in the previous chapter but in no way diminishes its probabilistic character.

it is not maladaptive—that is, that it is more or less neutral. But that is not enough. Explaining the success of our species requires acknowledging that we possess belief systems which not only are good at preventing us from engaging in activities which are directly harmful to ourselves but which also enable us to transcend an insipid existence of rote reflexive responses to the environment. Consequently, we are relatively good at believing things that are in our best interest to believe and the occasional newsworthy exceptions are remarkable precisely because they illustrate a normally useful system gone bad. If our belief systems were not largely reliable, we would not survive very long. This is not to say that we cannot harbor defective beliefs but that those beliefs must either be inconsequential or else be about things so far in the future that they have little or no immediate impact on us (even though they may eventually be of the utmost importance).

Faith is adaptive, therefore, because it makes life possible in an epistemological vacuum. We inhabit a universe in which truth, even at its most obvious, must always be inferred. As I've tried to show, there is no such thing as a choice between fact and faith—facts are merely those things to which we assign high probabilities. But faith is not only about what we believe with respect to the past and present but what we believe concerning the future. There is, therefore, a decidedly predictive nature to faith and, if faith is to be adaptive, those predictive probabilities must be relatively accurate. If they are not, we won't be bothered by them (or anything else) for long.

Neither the caveman's child (nor the hapless camper) who fails to properly predict the speed and agility of a threatening bear nor the pedestrian youth who misjudges how fast traffic is approaching a crosswalk are likely to pass their poor predictive skills to the next generation. Whether it is the simple predictive task faced by a hiker assessing the consequences of eating a certain mushroom or complex cognitive ones such as that faced by a general attempting to judge the strength of an opponent or a farmer trying to determine what crop to plant and when to plant it, all require functional predictive belief systems. But their predictions are merely what they believe to be true about the future (including how their actions will affect that future) and those individuals who can make the better predictions have a distinct survival advantage. None of this is to suggest that we are born with full-blown predictive skills—that, it appears, is intimately connected to our ability to learn—but that we possess brains which make relatively accurate predictions possible.[25]

---

25. See for example, Hawkins and Blakeslee (*On Intelligence*) and Donaldson ("Neural Network for Creative Behavior") on the efficacy of predictive learning systems. As Pinker notes, "A probabilistic inference is a prediction today based on frequencies gathered yesterday" (*How the Mind Works*, 351).

The spin-off of having prediction systems that keep us from harm's way is an ability to envisage the results of potential choices that increase our productivity and add to our happiness and sense of fulfillment. These are positive benefits that go well beyond mere survival advantage and include an aptitude for appraising the merits of accepting or declining a job offer or proposal of marriage, forecasting the reception for a contemplated work of art or literature, or projecting how any other of the innumerable plans we routinely make will unfold, all based on what we believe will be the likely outcomes. Such beliefs are our faith.

Because faith is adaptive, it makes no sense to disparage it per se. Any adaptive mechanism can be the source of potential problems but, if it remains adaptive (and faith does), then it cannot be essentially bad or it would not have been nor continue to be adaptive. It might seem easy at this point to counter such a claim by pointing to all of the other species who are imagined to have survived quite well without faith. But if faith is adaptive in the sense of having the predictive quality we have been considering, then we must ask if other species do not also possess some measure of it. I doubt this is an idea that everyone will find attractive but it is merely another way of acknowledging that faith is the product of brains. Only a mystical view of faith can prevent our coming to terms with this claim.

This need not mean that human expressions of faith do not occur at significantly more sophisticated levels than that of other animals but it does suggest that at least some of the underlying biological mechanisms are really quite similar. The probabilistic, predictive character of faith elaborated here is operative in every species that must gauge, among other things, the threat of a presumed predator or benefit of a presumed meal. From the nest-building activity of birds to food-caching behavior in mammals we see evidence of preparation for an unknown but partially predictable future. Every time a squirrel jumps from one tree to another it is making a veritable leap of faith.[26] It would be easy to see this as nothing more than a metaphor, but that is not my intent—the associated beliefs are still creations of brains.

The philosopher Alvin Plantinga, who would have us believe that accurate signal processing is sufficient for adaptivity and that the truth content of one's beliefs is irrelevant, claims that:

1. Adaptive behavior requires accurate indicators
2. Indication is not belief
3. Accurate indication need not be accompanied by true belief

---

26. Readers of the book of Proverbs are even instructed to consider the preparations of the ant (Prov 6:6–8).

4. As long as the indication is accurate, the belief content can be anything whatever[27]

Well, Plantinga is right as far as he goes but he just doesn't go far enough. Although his position might make some sense with respect to a frog, he maintains that it applies to humans as well.[28] But, even if this is true for mindless behavior, it is certainly not an adequate description of what happens in more sophisticated cognitive environments in general. Belief is built on indicators (i.e., the observational foundations we discussed in the previous chapter) and whether the indicators are believed or not can definitely make a difference in what is done by any being that is capable of reflection.

But who are such beings? From an evolutionary perspective one should not be surprised by a gradual shift from mere indication to full-blown belief. In fact, it is hard to see how belief systems could evolve in the absence of indicators, especially since beliefs themselves also serve in that capacity. Consequently, we might be wary of prematurely drawing too rigid a distinction between accurate indicators and belief content and especially between the conscious experience and potential belief content of other creatures and ourselves.[29] Do dogs, for instance, have beliefs (and not just the predictive character of faith described above for other species)? What about dolphins or chimpanzees? Some ethologists would probably say they do.[30] But if so, does it not matter if those beliefs are accurate? Is it appropriate to say that a dolphin constructing a silt fish-trap "believes" the effort will produce a meal? And would it really make no difference if the "belief" was incorrect?

Because he considers language essential to the possession of human-like concepts—that it is central to a genuine reflective ability—the philosopher Daniel Dennett would be reluctant to attribute human-like belief to dogs.[31] Whether or not language is the pivotal factor (and it probably is[32]), there is little reason to disagree but the question is not about belief identity between species but rather about the development of belief content across species. Dennett also claims that, "Mental contents become conscious . . . by winning the competitions against other mental contents for domination in the control of behavior, and hence for achieving long-lasting effects . . ."[33]

---

27. Dennett and Plantinga, *Science and Religion*, 66–70.
28. Ibid., 69–70.
29. Cf. the imagined consciousness hierarchy in Hofstadter, *Strange Loop*, 19.
30. Cf. Byrne, *Thinking Ape*.
31. Dennett, *Kinds of Minds*.
32. Cf. Bickerton, *Language and Human Behavior* and MacPhail, *Evolution of Consciousness*.
33. Dennett, *Kinds of Minds*, 155.

This certainly seems to describe the clever dolphin's apparently purposeful behavior and should make us reconsider any doubts we might harbor about such animals' potential for genuine beliefs, even if they are not quite human. However, because our well-developed reflective abilities render us competent to exceed mere stimulus-response prediction, we are accustomed to think of faith in its more cerebral manifestations. Yet even human consciousness waxes and wanes and there is little reason to assume that we consciously control all our decisions.[34] Perhaps we should be reluctant, therefore, to declare that all belief must be conscious.

In any case, it is our reflective ability that moves much of our faith beyond the cut-and-dried realm of rote responsiveness[35] and binary alternatives. Hume claims that our understanding of cause and effect are the result of experience and notes that one could not predict from the properties of a substance alone what to expect from it under certain conditions unless one had previously observed those effects: "Adam, though his rational faculties be supposed, at the very first, entirely perfect, could not have inferred from the fluidity and transparency of water that it would suffocate him, or from the light and warmth of fire that it would consume him."[36] Although I am in complete agreement with Hume regarding the necessity of experience, he doesn't give enough credit to the human ability to extend observation to new and as yet unexperienced realms. Had he known more of chemistry and physics, he could not have made such a statement, for the history of science since his day has done just what he said we cannot do—that is, make predictions about the behavior of things prior to obtaining the experience of them. As philosopher Richard Gregory aptly notes, "... perceptions in many ways are like predictive hypotheses of science. Both predict unsensed properties of objects, fill gaps in data, and both have limited prediction into future time."[37]

But the ability to transcend direct experience exists precisely because our predictive brains are capable of deep reflection. As Hume knew, even ordinary predictions take us beyond immediate experience, as they must in any world that is constantly changing and which involves other persons who

---

34. Recall McNamara's take on unconscious influences in *Evolution, Culture, and Consciousness*. Even when consciousness is involved, evidence indicates that at least in certain cases our brains may have already decided prior to our awareness what they will do (cf. Benjamin Libet, "Do We Have Free Will?" in Libet et al., *Volitional Brain*, 47–57 and Soon et al., "Unconscious Determinants of Free Decisions").

35. What Hofstadter has called "sphexishness" after the stereotyped behavior of a particular insect (*Metamagical Themas*, 529).

36. Hume, *Human Understanding*, section IV, 50–51.

37. Gregory, "Flagging the Present with Qualia," in Rose, *Brains to Consciousness*, 200.

are also trying to maximize their opportunities. When it comes to matters with significant long-term consequences, the prediction can be quite sophisticated. A student attempting to determine what to do after graduation, for instance, is likely to have little or no personal experience with any of the myriad options before her (Figure 2) but must rely on a brain conditioned by observation and conversation, including countless hours of television, movies and, hopefully, a few books. For each choice she must project herself into an imagined scenario and predict the outcome. The ultimate decision will represent her faith that one option is superior to the others. She may even be able to articulate the reasons.[38]

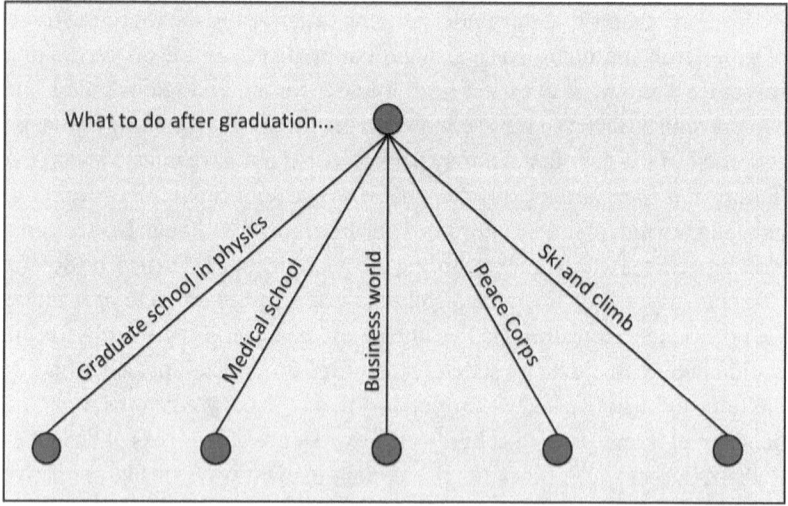

Figure 2. The predictive role of faith involving sophisticated cognitive capacities

In this case, faith is the prediction of possible futures and entails an evaluation of the likelihood of success or failure, happiness or sadness, fulfillment or emptiness, interest or boredom, fame or ignominy, wealth or poverty, health or infirmity, and so forth for each of the various options considered.[39] Words the student might use to describe her thinking on these issues—contemplation, deliberation, consideration, estimation, anticipation—are all faith words that reflect the probabilistic nature of her pending decision as well as each of the factors that can potentially influence it. Her eventual

---

38. Emotional factors that may be impossible to describe explicitly can play a critical role.

39. These considerations serve as constraints that prune the space of possibilities to manageable size.

choice will be the product of teamwork by millions of neurons, each influenced by and influencing its peers in a probabilistic dance that can scarcely be imagined by the student in whose head it is occurring.[40] Hofstadter likens the result to "a voting process in a democracy."[41] In any case, adaptation here means more than simple survival—it means survival with style.

But faith is involved in more than merely identifying the best course of action to pursue. Its predictive nature also plays a crucial role in actually moving individuals from their current physical state to the next, desired state. Without this aspect of faith, one would simply stagnate. Belief that the probability of a favorable outcome is high, therefore, can serve as sufficient motivation to engage in an activity that would otherwise be ignored or shunned.[42] For the cognitive activity of our hypothetical graduate to have any real significance, she must put her plans into action and it is precisely her faith in the outcomes of such action that motivates her beyond what would otherwise be a purely intellectual experience.

This motivating or enabling characteristic of faith provides the basis for venturing beyond what is treated as known and thereby extending one's perceptual and material horizons. Without believing in the merits of any particular venture, none would be tried and, even though some of those steps may be made in the absence of convincing probabilities (e.g., what we would better characterize as wishful thinking), without the proverbial mustard seed of faith[43] little of consequence could ever happen. As William James so aptly put it,

> Who gains promotions, boons, appointments, but the man in whose life they are seen to play the part of live hypotheses, who discounts them, sacrifices other things for their sake before they have come, and takes risks for them in advance? His faith acts on the powers above him as a claim, and creates its own verification.[44]

40. This may sound as though it is something happening to the student rather than a process which she orchestrates but it is too simplistic to reject freedom on that basis (e.g., consider Barbour, *Religion and Science*, 187).

41. Hofstadter, *Strange Loop*, 341.

42. Conversely, faith might just as well prevent the pursuit of some course of action deemed harmful.

43. Cf. Matt 17:20; Luke 17:6.

44. James, *Will to Believe*, 24. The essence of this idea is also conveyed by the observation that, "Faith is more than thinking something is true. Faith is thinking something is true to the extent that we act on it" (attributed to W. T. Purkiser in Maxwell and Dornan, *Becoming a Person of Influence*, 65). Ross succinctly notes that, "individuals behave as they believe" (*More Than a Theory*, 14). Experiments with brain-damaged patients, however, have demonstrated that there can be a fine line between knowing the

The world we inhabit is only apprehended by acts of faith—it is the only way we ever got to the world of today from that of yesterday and the only way we will get to the world of tomorrow if it is to be any different from the one we presently think we know. Perhaps this is the insight contained in the common phrase, "stepping out on faith." Though not without its risks, this, too, is adaptive.

Got Soul?

The mind-body problem[45] is concerned with how to understand the mental in terms of the physical. How are cognitive abilities to be explained as the products of brains? For many facets of human information processing this has become quite clear, and there is ample evidence indicating in great detail how brain chemistry and neural processing subserve behavior or support further cerebral activity. Certain competencies, however, remain elusive. For example, certain aspects of consciousness present particularly thorny dilemmas and, although there is an assortment of theories and models for symbolic processing and language, capacities such as these are still poorly understood. Attempts to unravel the details are met with various responses.

Just Do It

For many it is a foregone conclusion that the brain will eventually yield all its mysteries because of the assumption that a purely physical explanation exists and merely has to be found. The question for proponents of this view is not "Can this really be?" but "How does it work?" There is thus no fundamental mind-body problem. Many, if not most, neuroscientists, cognitive psychologists, philosophers of mind, and artificial intelligence researchers probably share this view.[46] The artificial intelligence contingent is particularly revealing in that the premise for its efforts resides in the belief that human artifacts can (eventually) have most, if not all, of the features of human cognition.[47]

---

odds that something is beneficial or harmful and actually acting appropriately on that knowledge (Cf. Bechara et al., "Emotion").

45. This is the legacy of Rene Descartes but we know it concerned Plato and, therefore, most likely many before him. See Peter Harrison, "René Descartes," in Numbers, *Galileo Goes to Jail*, 107–14 for a refutation that Descartes was the first.

46. Cf. Crick, *Astonishing Hypothesis*; Jaynes, *Origin of Consciousness*; Dennett, *Sweet Dreams*.

47. One of the best known early articulations of this hope was by Turing ("Computational Machinery and Intelligence"). McCulloch and Pitts ("A Logical Calculus")

Yes, But . . .

Some scientists on the physicalist bandwagon, however, are still hedging their bets but for presumed scientific reasons (e.g., non-computability[48]—a technical term I'll discuss in the next chapter.) Other individuals are prepared to qualify our understanding even before we have achieved it. Philosopher John Searle, for instance, identifies certain computational approaches to mind as incompatible with human cognition.[49] Although probably more optimistic than Searle, famed mathematician John von Neumann considered it "perfectly possible that the simplest and only practical way actually to say what constitutes a visual analogy consists in giving a description of the connections of the visual brain."[50] Philosopher David Chalmers suggests that, "conscious experience is just not the kind of thing that a wholly reductive account could succeed in explaining."[51]

Trying to Hold On

Then there are those who live in apprehension of a physical explanation for other reasons.[52] This may be the position of some readers who will have trouble with the claim that faith is a natural product of brains and would like to see it treated, instead, as some mystical phenomenon. But faith is not mysterious, even though there may be plenty of mystery about the objects of one's faith (and possibly about the reasons for it). It is one thing to claim that faith itself is a product of physical brains in this world and something altogether different to suggest the possibility that certain instantiations of faith may have otherworldly implications.[53]

---

had previously shown that neural behavior can be characterized as the operation of binary logic devices. The literature is vast but, for some interesting perspectives, Albus, "Outline for a Theory of Intelligence" and Moravec, *Robot* are instructive places to start.

48. Cf. Penrose, *Emperor's New Mind*.

49. Searle, "Minds, Brains, and Programs."

50. John von Neumann, "General and Logical Theory of Automata" in Jeffress, *Cerebral Mechanisms*, 24.

51. Chalmers "Hard Problem of Consciousness," in Velmans and Schneider, *Companion to Consciousness*, 234.

52. Cf. Thomas Clark, "Fear of Mechanism" in Libet et al., *Volitional Brain*, 279–93.

53. You don't have to assume an evolutionary origin for faith to accept that it is a product of the brain, but evolution is often thought to pose a significance-reducing threat to humans. Drummond, *Ascent of Man* and Teilhard de Chardin, *Phenomenon of Man* have suggested the opposite.

Theologian John Haught suggests that people find comfort in mystery[54] but I don't think that is quite right. Experience indicates that people enjoy mystery precisely because of the potential for its resolution. We all like a good mystery but not if it remains a mystery. Imagine how ultimately disappointed you are with those occasional books or movies that keep you guessing right up to—but then beyond—the end. They are enjoyable as long as you believe you will eventually understand but anticipation quickly gives way to disenchantment with the discovery that you will be forever in the dark.

In most, if not all cases, mystery without hope for solution is comforting only to those with no curiosity and enjoyable only by the lazy or forgetful. Furthermore, demystifying faith will not automatically suck the life out of all further mysteries but it may help solve a few. This need not frighten us, however. There are many things we might legitimately fear (including failure to resolve some important mysteries) but loss of all mystery is not one of them.[55] What is mysterious is why any individual with theistic leanings would be very concerned about this. The inestimable attributes of deity would seem to imply an endless supply of problems to be solved, meaning the devout should actually be the least concerned about the loss of mystery.

Lack of concern, however, is sometimes the result of having prematurely drawn the line indicating what we might or might not have the wherewithal to fathom. I imagine there are such lines but I'm more sure that the person who draws them is doomed never to move beyond while the less inhibited forge ahead.

In any event, these line-drawing types of presumptions are not limited to the religious. Science writer John Horgan's proclamation that all significant theories have now been produced[56] is a purely secular lament about the loss of mystery accompanying the rise of modern science. Physicist David Lindley is less pessimistic (if for no other reason than he only deals with one science) but still argues that physical theory is rapidly approaching the point of stagnation at both cosmological and sub-atomic levels.[57] This is not the place to rebut such outlooks, as effective written responses have already been given[58] and each major advance in the future will be even more telling. Although proclamations about such loss seem rash, presumption (as we are about to see) can run deeper.

54. Haught, *Christianity and Science*.

55. As the physicist Paul Davies claims, "There will always be mystery at the end of the universe" (*Mind of God*, 226).

56. Horgan, *End of Science*.

57. Lindley, *End of Physics*. See also Stannard, *End of Discovery*.

58. For a summary of Horgan's views and some excellent responses see Horgan, "Why I Think Science is Ending".

What About the Soul?

The prospect of interpreting the mental in terms of the physical can appear to effectively shut the door on the possibility that the mind could be influenced in a non-physical way. This, I think, is the real dilemma for many. The mind-body problem is actually perceived as a subset of a more profound soul-body problem (without the necessity of a religious connotation although there usually is). Consequently, there is substantial resistance to and even fear about the prospects of the reductionist program.

Attitudes toward the soul, of course, range from terse denial that it even exists to adamant insistence that it is the very essence of who we are. Intermediate views merely equate it with consciousness[59] or take some other compromising position. Although it is easy to pay lip-service to one side of the debate or another, I'm willing to bet that most folks haven't moved far beyond the shouting stage.

One way to help you see where you come down with respect to this issue is to construct a Venn diagram using the labels "soul, mind, consciousness, and brain" for each of the subsets. Figure 3 illustrates a few examples of possible relationships between these entities.

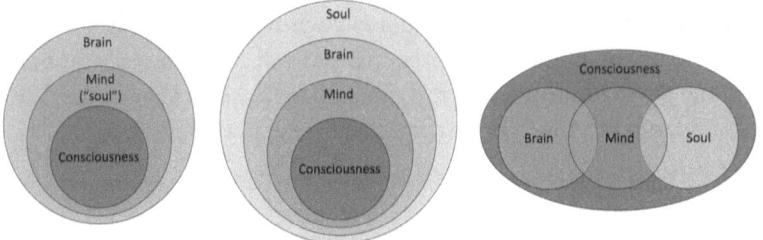

Figure 3. Some possible relationships between "soul, mind, consciousness, and brain"

Construction of your diagram will obviously depend upon how you define each of the terms but also on whether you intend it to be a physical or metaphorical representation. For instance, a drawing in which "soul" subsumes "brain" might be taken physically to suggest that the brain is but a temporary repository of some portion of the soul or metaphorically to suggest that the soul is more significant than the brain. In any case, there is no requirement that your sketch look like any of those above. If you are the strictest of materialists, for instance, you might simply draw a single circle to represent all four components. I really don't expect you to show

---

59. Cf. Hofstadter, *Strange Loop*.

four non-intersecting circles unless you've been devouring a steady diet of sci-fi but, if that reflects your best assessment, go for it. What I do expect is that you should be able to justify your diagram according to some reasonable criteria.

Once you've settled on a representation that reflects your understanding, pencil in a single dot for the element "faith" to show how you believe it relates to the four sets. Finally, contemplate the implications of your work. For example, a dot (for "faith") positioned in the rightmost diagram in Figure 3 at the intersection of "Mind" and "Soul" would suggest that faith is a property of both but that it is not something associated with the brain. Frankly, I hope you didn't do that! Doing so would indicate that what we've considered in this chapter so far hasn't had much impact on your thinking. On the other hand, a dot inside the "Consciousness" set in the leftmost diagram in Figure 3 would indicate that faith is also a property of the brain (because "Consciousness" is a subset of "Mind/Soul" which is a subset of "Brain").

My concern here is not to argue against the soul but to challenge poor concepts of it[60] which get in the way of a productive understanding of faith. Recognizing that there is a physical substrate for faith is probably the most important step away from the pernicious view that faith is just a religious phenomenon. Faith may have implications for souls but it is not confined to them. Neither do we have to omit the soul when giving the brain its due.[61]

I harbor no illusions, however, that this will go down easily for everyone. The physicist Stephen Barr, arguing against a strict materialist view of the mind, submits the following argument in support of his case:

> If ideas are just patterns of nerve impulses, then how can one say that any idea (including the idea of materialism itself) is superior to any other? One pattern of nerve impulses cannot be truer or less true than another pattern, any more than a toothache can be truer or less true than another toothache.[62]

Now this is not an uncommon argument but Barr's example actually shows why a materialist position is so convincing to many. As it turns out, one toothache *can* be "truer" than another. Contrast, for example, a toothache generated by an unhealthy tooth with one caused by electrical stimulation of the appropriate area in cortex. Both potentially involve the same set

---

60. Such poor concepts include giving a nebulous soul too much explanatory power in our physical universe or postulating a straw man version of the soul which any ten-year-old could attack (Cf. Minsky, *Society of Mind*, for an example).

61. If too much emphasis on the brain appears to put prospects for immortality in jeopardy, see Green, *Body, Soul, and Human Life*.

62. Barr, *Modern Physics and Ancient Faith*, 197.

of neurons and from the individual's perspective may appear identical (i.e., I have a toothache), but in one case the belief that a tooth is the culprit is true and in the other, false. Anyone who has given much thought to virtual reality will immediately recognize the point: Is my experience and corresponding belief content real or merely simulated?[63] In any case, the mere ability to stimulate a particular area of your brain and make you believe a particular thing (i.e., that you have a toothache) should be sufficient to demonstrate that it is the brain that is mediating belief formation.

If we are able to escape the hope, hype, and hysteria that often accompany consideration of the soul-mind-consciousness-brain problem, it is hard to deny that all those features we associate with who we are—our personalities, behaviors, emotions, thoughts, and memories—depend on the brain. If you doubt this, even for a moment, consider the mind-changing impact of drugs, sleep-deprivation, trauma, aging, and hypnosis. Or, ponder what neurological and psychological research have revealed about the physical substrate as it pertains to the mental disposition of individuals suffering from assorted agnosias, autism, or Alzheimer's (or even an affinity for alliteration). We are at a loss to explain belief in the absence of a brain, but we do know that when even a part of the brain is absent (e.g., due to dementia, stroke, or other trauma), beliefs can be dramatically affected.[64]

In short, physical changes to the brain have consequences for the mind. Extrapolating from the implications for mentality due to alterations in brain structure or chemistry leads to the conclusion that no brain means no mentality. It also means no faith. I hope this is beginning to strike you as tautological. On the other hand, acknowledging the brain's role in matters of faith is not incompatible with a deeper perspective of what it means to be human. I think Barr's big concern is with what Keith Ward critiques as a "nothing more" perspective—that there is nothing else to us beyond that which is describable in terms of physics or chemistry.[65] As the linguist Ray Jackendoff reminds us, our understanding of the role of genetics and brain structure are articles of faith.[66] However, while the faith Jackendoff

---

63. Cf. Schneider, *Science Fiction and Philosophy*.

64. Cf. Sacks, *Man Who Mistook His Wife for a Hat*.

65. Ward, *Big Questions*, 142–45. The list of features I mentioned in the previous paragraph (i.e., those we associate with who we are) are quite similar to those used by Crick (*Astonishing Hypothesis*, 3) in making one of those "nothing more" claims with which Ward (143) takes issue. Although stressing the crucial role of the brain, I make no such claim. As Robinson notes, "If the mind is the activity of the brain, this means only that the brain is capable of such lofty and astonishing things that their expression has been given the names mind, and soul, and spirit" (*Absence of Mind*, 112).

66. Jackendoff, *Consciousness and the Computational Mind*, 88.

describes is well supported by scientific observation, the faith that there is "nothing more" is not.

* * *

Not only is it unnecessary to postulate a metaphysical basis for faith, doing so significantly reduces the chances for obtaining an accurate and useful understanding of the concept. Furthermore, we must never forget that there is a distinct difference between faith and its objects. Faith is a probability assigning operation embodied in the brain—we see its effects every day. Those things about which it makes judgments, however, can be real or imagined.

PART II

Faith and Reason

── CHAPTER 6 ──

The End of Knowledge: A Challenge for Faith

> Lord, make me to know mine end, and the measure of my days, what it is: that I may know how frail I am.
>
> PSALM 39:4[1]

AT SOME TIME IN your life you may have wondered what it would be like to have a brain like Newton's or Einstein's or some other notable genius. The truth is, you may actually have one but never know it. We judge such brains by the accomplishments of their owners but any comparable lack of achievement on your part doesn't necessarily indicate inferior intelligence. Both chance and just being in the right place at the right time are also major contributors to scientific creativity.[2] In addition, when we look at the actual lives of creative individuals we discover that they were subject to most of the same foibles and limitations we observe in our own. If you're now feeling pretty good about yourself, permit me to further bolster your smugness by pointing out that you probably know many things today those luminaries scarcely imagined.

It is not my intention to belittle the contributions—indeed, they are envious—of those we call great, but merely to provide a friendly reminder that omniscience is something typically reserved for deity. Trying to know

---

1. King James Version.
2. Simonton, *Creativity in Science*.

everything (with the highest degree of certainty) is a laudable objective, but any belief that such a vision can be completely fulfilled is a pipe dream. The only recourse is faith. In other words, our inabilities to know everything and to possess whatever knowledge we do with complete confidence are just what precipitate the need for faith, even for the gifted.

Faith, then, is both a function of and dependent upon our impediments but, as Dirty Harry was kind enough to remind us, "A man's got to know his limitations."[3] Actually, it is important for all of us (man, woman, or child) to know this, so, despite Harry's cinematic chauvinism, let's pretend that he meant it in a gender-neutral way. In this chapter, then, we will consider the limitations that make faith necessary:

Limitations that Necessitate Faith

1. Human brains are physically constrained.
2. Human beings are chronologically constrained.
3. Human brains are logically suspect.
4. Answers to many significant questions are limited to approximations.
5. Certain types of measurements are always approximations.
6. Theories are just approximations.
7. Even the most precise formalisms have built-in limitations.
8. The more you know, the more I must accept on faith.

Like it or not, these limitations are inescapable and provide reasons for our reluctance to claim that we *know* anything. With sufficient effort, however, we can make headway against some of them, as we shall see.

Human Brains Are Physically Constrained

As we saw in the last chapter, mental competence is enabled by the signaling of neurons in one's brain. According to current estimates,[4] you and the best Trivial Pursuit player alive today probably each have less than 100 billion of those special cells, every one potentially connected to thousands of others. Each neuron is able to receive and transmit signals at the rate of thousands per second. In many respects this is quite impressive but, the truth is, those

---

3. Post, *Magnum Force*.
4. Cf. Herculano-Houzel, "Human Brain in Numbers."

numbers are relatively fixed and, by some standards, looking smaller all the time. Signaling in computers, for example, takes place thousands to millions of times faster than in brains. In fact, overall storage capacity and processing speeds in digital computers have been zeroing in on comparable human performance for a number of years.[5]

Finite brains mean that, despite the seemingly inexhaustible supply of long-term memory, it is really a limited commodity. Short-term memory is far more constrained.[6] For example, trying to recall several new telephone numbers or names is problematic for most of us, especially when engaged in other tasks.[7] Even after we think something has been permanently learned, it is often disappointing to notice that we are unable to retrieve it.

The learning process itself depends upon input from fragile and fallible sensory devices (eyes, ears, etc.) that are sometimes augmented with aids in an attempt to improve their efficiency but, as we've already seen, we can only go so far. Granted perfect vision, we would still be restricted to observing a maximum of three spatial dimensions and can only imagine more, perhaps mathematically or as a thought experiment (although we bog down quickly when we try to do so).[8]

I don't mean to be glum, but your brain and mine are limited processing devices. Even if efforts to artificially expand our cognitive abilities[9] are successful, they will only extend the limits—they will not erase them. Moreover, everything I have said so far refers to the limitations of healthy, normal brains. Things are worse—perhaps much worse—when our brains are not at their best.

In 1973 a group attempting to climb Aconcagua, the highest peak in the western hemisphere, met with disaster when two members of the expedition were killed and several others suffered various forms of altitude related injury. Two of the climbers

> arrived back in base camp still hallucinating. While above 21,000 feet they had seen mules, pieces of heavy road moving equipment and skiers. Until their arrival back in base camp, they *felt secure in the belief* that their missing companions had

---

5. Moravec, "Computer Hardware."
6. Cf. Miller, "Magical Number Seven."
7. Cf. Baddeley, "Working Memory."
8. Cf. Abbott, *Flatland*.
9. Cf. Roco and Bainbridge, *Converging Technologies*; Naam, *More Than Human*; Garreau, *Radical Evolution*.

been rescued by the Argentine mountain patrol—whose presence and voices they hallucinated.[10]

Although what the climbers thought they had seen may not seem unusual for tourists in the Sierras, it was impossible to justify in Aconcagua's wilderness setting and at such an altitude. Furthermore, although they showed high confidence regarding their teammates' safety, it was entirely misplaced.

This is the kind of mental activity one expects from substance abuse but even a natural high can play havoc with accurate belief formation. Yet, getting (physically) high is not the only way this occurs. SCUBA divers know that increasing the depth of a dive can result in hallucinations accompanying changes in the partial pressure of gases in their tanks—a phenomenon euphemistically referred to as Martini's Law. The point of these examples is that the creation of appropriate beliefs is severely compromised when the brains in which those beliefs develop are subjected to environmental influences occurring outside an extremely narrow range of acceptable values. You may feel secure if you are not a climber or diver but try to assess the clarity of your thinking the next time you are huffing and puffing through a morning workout or have allowed your blood sugar to get a bit off-kilter.

Such molecular fences that confine our otherwise healthy but inevitably fragile brains, however, are relatively minor compared to those of a more permanent nature. In particular, a host of potential neural deficits threatens to further restrict our cognitive capabilities. Cortical blindness, schizophrenia, various forms of retardation—all these and many, many more difficulties with cerebral function lead to noticeable restrictions in thought or behavior. The most pernicious of these deficits but, in some respects, the most illuminating fall into a category called anosognosias.[11] These are deficits (perhaps of the visual or mental imagery systems) accompanied by failure on the part of patients to recognize that they have a problem.[12]

Hemispatial neglect is one of these that has received a great deal of attention from neurologists. It can occur when (for example) a stroke to an area in the right parietal cortex leads to an inability to perceive the left half of one's visual space. Individuals with the corresponding lesion will omit significant portions of the left side of drawings they are asked to copy or will fail to describe the left half of what would normally be a familiar scene they

---

10. Schults and Swan, "Retinal Hemorrhages," 19–26 (emphasis mine).

11. Cf. Heilman, "Anosognosia," in Prigatano and Schacter, *Awareness of Deficit*, 53–62.

12. Ibid., 53.

have been requested to visualize.[13] What makes this especially bizarre is the apparent inability of some individuals to appreciate that they are performing abnormally.[14]

But "normal" is only relative to what a majority of us believe to be true about ourselves. If you have neglect, you will appear subpar to those around you but you may not know it. I trust that is not the case, but those times we have spent insensitive to an unzipped fly or unbuttoned blouse, a dryer sheet dangling from a sleeve, halitosis, the real significance of an offhand remark, or a self-conception shared by no one but ourselves are reminders that analogs to neglect span a broad continuum. It is a bit sobering to consider the likelihood that, somewhere between the trivial and the devastating, there are aspects of our day-to-day physical or mental environments—some with potentially life-changing possibilities—to which we are completely oblivious. Because finite brains must always have limited awareness, there will necessarily be some things about which we can never know. But brains constrained not to know everything have no recourse but to faith—to probabilistic assessment even in those areas that can impinge on awareness.

Human Beings Are Chronologically Constrained

The Wall Chart of World History[15] is a lavishly illustrated guide to a variety of significant events and persons spanning 6000 years of human history. Perhaps you own or have seen one. The pages of the book fold out to a length of some thirteen feet to depict time as distance. As you might guess from the time frame represented, it begins with a date corresponding to a biblical chronology. The timeline necessarily stops short of the present since its publication date is continually slipping further into the past. A drawing of Adam and Eve grace the leftmost end of the chart and a computer the rightmost. Although skipping more than four billion years of earth history, quite a bit seems to have transpired between those times.

I mention all this because it is difficult to look at the chart without thinking about the shortness of human life relative to the duration of recorded human history, much less the life of the cosmos. Martin Luther, for instance, gets about one and three-quarters of an inch in the chart for his

---

13. Kandel and Kupfermann, "From Nerve Cells to Cognition," in Kandel et al., *Essentials of Neural Science*, 321–46.

14. Cf. ibid., and Heilman, "Anosognosia: Possible Neuropsychological Mechanisms," in Prigatano and Schacter, *Awareness of Deficit*.

15. Hull, *Wall Chart*.

sixty-three years on this planet; Galileo about two inches. It's not yet clear how much you or I would get but it is apparent that, relative to everything that has transpired on earth, we get to experience an exceedingly small fraction of it. Barring significant medical breakthroughs, the portion for our progeny is smaller still.

Consequently, being confined to a specific moment of time has significant implications with respect to the role of faith in our lives. The inability to live during all eras of time limits our capacity for primary observation, meaning the potential for acquiring direct knowledge of entities—Edward III, Edison, or Edsels, for example—is constantly being lost to the past. As a result, many things we call "facts" are really just what we believe to be true based on someone else's report. Our incapacity to exist at all times, of course, is complemented by a geographic constraint which places comparable limitations on our primary experience. In both cases, relative to what could be known firsthand by an omnipresent, omnitemporal being, we can personally verify only a smidgen.

But life is not only short in a relative sense. The time we can devote to acquiring information is restricted to a few years of existence and there is only so much we can read or watch during that period. This means that the confidence placed in any particular belief is not only bounded by the credibility assigned to sources but that even the process of assigning credibility is constrained by the time available to establish it. In addition, the more time we spend trying to bolster our confidence in one thing the less time there is for another.

Human Brains Are Logically Suspect

"It's a bald eagle!" The cry rang over the base area as we prepared to board the chair lift on our way to the first ski run of the day. Practically everyone within earshot turned their gaze skyward, only to discover as the bird drew nearer that it was merely a raven carrying a piece of white trash in its beak. How in the world could a national emblem be mistaken for the symbol of Poe's distress?[16] But is there really much chance of getting through life without such gaffes? Can we commit, nevermore, to leap to erroneous conclusions, repeat our mistakes, or otherwise appear foolish?

Replacing one person's eagle belief with another's raven belief may look like progress but I'm no ornithologist. Was it really a raven? Had we not been in the Rockies, I might have called it a crow. Had I never seen raven or crow, I might have called it a sparrow—relatively safe, but hardly definitive.

16. Poe, "The Raven."

A number of authors have spent a fair amount of energy pondering the frailty in our thinking processes.[17] Although most of us are smart enough not to need someone else telling us that we are prone to make dumb mistakes or that not all of our beliefs are sound, it is instructive to contemplate some of the ways in which this occurs and the reasons why.

Muddled Thinking

Much insight into this phenomenon has come from psychologists Amos Tversky and Daniel Kahneman[18] and it is not uncommon for people writing about our foibles to refer to pioneering work done by them or in conjunction with them. I first heard Tversky speak on this subject when I was in graduate school, shortly before he died of cancer. One of his illustrations in that talk involved the so-called "hot hand" in basketball. You've probably heard the expression used to refer to a player who seems to be on a roll, the assumption being that such a player is more likely to make his next shot because of his current streak. Unless you're familiar with the work of Tversky or his colleagues, it may surprise you (and a lot of coaches) to discover that there really is no such thing. A careful study of two professional basketball teams as well as college players provided empirical evidence that the "hot hand" is a myth.[19] Nevertheless, I still recall my basketball-loving advisor leaving the talk questioning Tversky's assessment. I am sure that, as a scientist, he felt compelled to go with the evidence but his "common sense" told him otherwise. That, I believe, was Tversky's point.

In his book *How We Know What Isn't So*, social psychologist Thomas Gilovich (who was first author in the "hot hand" study) provides a number of reasons for our propensity to form less than valid judgments. Chapters in his book are devoted to (among other things) explaining how we misinterpret random data, infer too much from inadequate information, see what we want or expect to see, are biased by hearsay, and envision more support from others than is warranted.[20]

The essence of this message is certainly not new. In the early 1600s Francis Bacon described the "idols of the mind"[21]—those inborn or socially

---

17. Cf. Gilovich, *How We Know*; Mlodinow, *Drunkard's Walk*.

18. Cf. Tversky and Kahneman, "Judgment Under Uncertainty" and "Framing of Decision."

19. Gilovich, et al., "Hot Hand in Basketball."

20. Gilovich, *How We Know*.

21. Bacon, *Great Instauration*, "Plan" and "Novum Organum," aphorisms XXXVIII–LXVIII.

derived predispositions and prejudices that shape our beliefs and, all too frequently, lead us astray—and philosophers and others have long sought to correct our tendencies toward muddled thinking. Unfortunately, such efforts must be repeated for each new generation and, in any event, have met with limited success. Breakdowns in our inductive and deductive thinking deserve a sizeable share of the blame for this state of affairs.

Much of what we call common sense is the product of inductive reasoning—coming to some general conclusion based on one or more observations. Inductive inferencing, therefore, might discourage us from crossing a stream where our jungle leader has just succumbed to a piranha attack, believing that we, too, would be eaten. However, although there can be significant survival advantage to this process, it is not without its failures, ranging from old wives' tales to all sorts of unjustifiable prejudices. Inductive reasoning can be considered logically sound only when the observations exhaust all possibilities, but this is frequently not possible. My inferred bias against crossing the piranha-infested river, for example, might appear reasonable, but how do I really know that they would eat me, too? Perhaps they are sated with the first victim, don't like hairy legs, etc. I am unable to test all possible combinations of piranha encounters if for no other reason than, if I am eaten, further tests are impossible.

When my seventh-grade daughter's middle school softball team won their opening game by fifteen runs, one might have made the inductive inference that they were headed for a great season. In fact, I was so inclined. Gilovich, however, would point out that this is an instance of inferring too much from inadequate information and that the conclusion is unwarranted.[22] That was easy enough to see when the team lost the next two games by a combined thirty-five runs but the admonition against over-generalizing remains valid even if the outcomes of those games had been reversed (although one would probably feel more secure about inferring a super season). Many of the reasoning blunders mentioned by Gilovich are the result of jumping to poor inductive conclusions.

Formal logic, on the other hand, is termed deductive and constitutes a rigorous process for drawing legitimate conclusions under the proper circumstances. Those circumstances include the availability of valid assumptions and correct application of deductive techniques. There are myriad ways in which to go astray when attempting to reason deductively and human brains are subject to all of them. The most pernicious, however, is failing to undertake the task in the first place. This could be the result of a conscious decision not to pursue something to its rational conclusion but

---

22. Gilovich, *How We Know*.

more often it is probably due to giving little thought to the possibility that a logical analysis could take one down a path of discovery. We'll have more to consider along these lines in the next chapter but, for now, I'll illustrate the problem with respect to syllogistic reasoning,[23] perhaps the simplest and best known application of forming deductive inferences.

Imagine a student named Tim who recently took a job in order to help pay for his college education. Unfortunately, Tim failed one of several courses taken during the semester in which he began working. Tim's objective is to earn his degree as quickly as possible, believing that doing so will be financially advantageous. Consequently, he is focused on the following rules:

multiple courses per semester → shorter time in college

shorter time in college → financial advantage

Simple syllogistic reasoning has thus led Tim to conclude that if he takes multiple courses each semester he will end up better off financially. Although the reasoning appears sound to Tim, it is problematic because it neglects to consider whether the rules themselves (his assumptions) are legitimate. In light of his recent failing grade, you might suspect that the first rule should be modified as:

multiple courses per semester and not fail courses → shorter time in college

If Tim's work load prevents him from devoting adequate time to his studies and if he therefore fails one or more of the courses he is trying to take each semester (another syllogism), he might very well be better off taking fewer courses. Repeating courses he has failed will both extend his time in college and increase tuition costs, thus impacting him financially on two fronts. Rules regarding the effect of course load on GPA and subsequent marketability can also be brought into play but this simple example should be sufficient to illustrate that even properly executed deductive techniques can lead one astray if the rules are improperly formed or assumptions are otherwise invalid.[24]

---

23. Syllogisms are just a chain of if-then rules applied successively. A rule can be written as A → B, read "A implies B" or "if A then B." If we know a collection of rules such as A → B, B → C, and C → D, and if we know A to be true, we can reason syllogistically (based on a technique called modus ponens) to infer that D is true (cf. Gersting, *Mathematical Structures*).

24. Short-term memory limitations can have an effect on syllogistic reasoning, as can failure to treat some rules probabilistically.

Invalid Assumptions

I suggested above that if we begin reasoning from a set of false premises, all bets are off as to the soundness of our conclusions. This is true in both the inductive and deductive worlds. We have already seen that inductive reasoning is not logically sound, but it is regularly utilized, nevertheless, to provide the assumptions upon which further reasoning proceeds. But nothing makes much sense when our premises are invalid, whether they were obtained inductively or otherwise.

Let me illustrate this by asking you to take a look at the photograph in Figure 1. Your task is to provide a caption for the picture. As you do this, keep tabs on the various descriptions that come to mind.

Figure 1. What caption would you suggest?

A large, framed print of this photograph is hanging in my home commemorating a kayak trip through the Grand Canyon. When my wife (also the photographer) had the photo framed, she was assisted by an acquaintance of ours who was employed at a framing shop near our house. I suspect that if I had asked that young lady for a caption at the time she did the work, she would have submitted something like, "Crazy man running a waterfall."

I only surmise this because at a later date the framing assistant was in our home and happened to observe the photo she had helped frame, hanging at that time above my desk. Studying it for a moment she finally asked with a puzzled expression, "Where's the water?" The "water" she wanted to know about wasn't the river in which I was about to land but the waterfall she had apparently decided I was running. Presumably, her assumption that a vertical kayak with cliffs in the background meant that the paddler was running a waterfall had caused her to imagine one that did not exist. When she finally noticed the absence of the inferred waterfall, the image no longer made sense.[25] But that is the problem with faulty assumptions—when we start with them, things just don't add up. Not noticing the errors, we then draw conclusions that become the assumptions for yet another round of mistake-laden thinking, all the while reveling in our impeccable logic.

Memory Anomalies

Faulty assumptions can result from slip-ups in any of our modes of observation but they can also arise from glitches in memory. This is a particularly insidious source of error because the very nature of the problem renders us unaware of it and makes it difficult even to gain such awareness. This is conceptually akin to anosognosia as described earlier but occurs in normal brains. Fortunately, our ability to lead meaningful lives suggests that many of these glitches are relatively trivial.

In chapter 5, I discussed a Galen Rowell photograph in order to illustrate how our brains can make decisions automatically and below our level of conscious awareness. When I first wrote the paragraph, my mental imagery system was flirting with an attempt to recall the picture to which Rowell referred and, based on the image I had in mind, I alluded to "the grass at the base of a precipitous mountain cirque." It had been a number of years since I had actually read the text and so I decided it might be wise to double-check the reference. In addition, I was interested in seeing once again the actual picture that had generated Rowell's observation. Needless to say, I was a bit

---

25. As you may have surmised, I merely launched my kayak off the cliff assuming that it would make a good picture.

shocked (but not entirely surprised) to discover that there was no cirque—in fact, there was no photograph at all! I had remembered his point fairly well but had fabricated an image that was not included in his book.

Tendencies to manufacture false memories often escape our notice but cognitive psychologists have been aware of it for decades. I do confess a slight embarrassment in admitting my error but couldn't have hoped for a much better illustration of how belief formation and our assessment of it are not always on the same page. In any case, I changed my original wording.

At times, however, memory malfunctions can have serious repercussions. A prosecutor or defender, for example, may build an entire case on the assumption that a witness is credible, leading a jury to believe with great confidence in the guilt or innocence of the defendant. Yet, if during the course of the trial the memory of the witness is shown to be flawed, the arguments come crashing down.

We've already considered the predicament accompanying traumatic brain damage but let's revisit that for a moment to get a better perspective on the potential influence of memory anomalies. For example, consider how an individual with retrograde amnesia (perhaps due to advanced Alzheimer's) would respond to the task we tried in chapter 4 of making a list of things known for sure. Items we take for granted such as where we live, who our family members are, and even our names might no longer be recalled and hence would be unavailable as the assumptions from which any reasoning process (leading to belief formation) could proceed. It is not likely that this describes you now but, because of the reconstructive nature of memory and the problems of forgetfulness, you fall somewhere between this and an impossibly perfect memory.

Although observational errors, muddled thinking, invalid assumptions, and memory anomalies combine in interrelated ways to produce beliefs about the world that sometimes wilt under the glare of serious scrutiny, those beliefs and the behaviors they engender represent who we are at our most fundamental levels. Fortunately, many of those difficulties are preventable or at least correctable. As we are about to see, however, there are other factors that will always limit us.

Answers to Many Significant Questions Are Limited to Approximations

If there is any doubt about the truth of this contention, taking exams and, especially, grading them, is a sure-fire way to become convinced of it. Most of us can recall (though we may not want to) some essay or arithmetic proof in which we were less than perfect in regurgitating the fullness of human knowledge on our assigned topic. Even those things we think we know best are usually no better than approximations. This may be little consolation when agonizing over a botched exam but maybe it will help to know that teachers are not immune. Your English professor's target for an analysis of Hamlet, for example, was a moving one, perhaps varying from student to student and semester to semester, despite her attempts to be objective.[26]

But approximation doesn't just happen on academic exams. We speculate and conjecture on tests of our judgment in all sorts of scenarios including the very intimate. I hope you will forgive my being a tad personal, but I have another little exercise for you that will make abundantly clear how even those things nearest and dearest can be viewed as approximations. To begin, think back to the time before you were married—if you are single, this will be easy!—and imagine all of the attributes you desired to be present in your future spouse. This will be your conception of Mr. or Ms. "Right." Your inventory will possibly include categories for physical appearance, mental health, character, financial stability, likes and dislikes, communication skills, and so forth, but this is just to get you thinking. I don't want to presume what should or should not be in your list—it's yours. However, if you are already married or engaged, I need to warn you that there may be a tendency to make your list of desired attributes fit your current spouse or fiancé. Despite the apparent safety in doing so, try to resist less you miss the significance of what can be learned from this experiment.

Once you've cataloged the desired attributes, try to recall the people you dated (or are dating) and arrange them on a number line according to how well they meet your overall criteria. Use any scale you like although, for consistency with our previous discussions of probability, you might select a range from 0.0–1.0 with 1.0 denoting a perfect match to all your desires. If someone on your list rated a 1.0 and if you are married to that person, congratulate yourself on your good fortune, then ponder where you would fit on his or her scale. You might also mull over the fit between your original and current perceptions and note that even the relative values you imagined for items in your list of attributes were only estimates.

You'll probably agree that it can be a bit sobering to consider your spouse as an approximation but, unless you believe you had no real choice in the matter—unless you are also wedded to an unshakable belief in fate—that's the way things are. This becomes even more obvious when you consider how many people are on your scale. For the sake of argument, let's say you dated 100 different people. That will seem like a large number for many

---

26. Cf. Mlodinow, *Drunkard's Walk*.

but the sample space is astronomically larger. Whether you were dating years ago or are currently doing so, that still leaves several billion members of the opposite sex for which you have not accounted.

When the number of alternatives is large, approximation is often the only recourse. Constraints of time or other resources can force us to accept a solution deemed good enough, knowing it may be sub-optimal. The technical term for measurements of the resources required to search a space of possible solutions is termed "complexity." Depending upon the problem at hand, those requirements can be few or vast. As noted above, for example, the search space of possible people to date is huge but we prune it to manageable proportions by dint of geographic, ideological, and other constraints. On the other hand, the search space of possible spouses among just those you have dated is obviously far smaller. If you have an arranged marriage, the space is smaller still.

To get a better handle on the role of approximation, particularly when dealing with large search spaces, let's consider what occurs when playing a game such as chess or checkers. Your objective is to make the best move at each turn but there are typically multiple possibilities. How do you decide? A nice way to visualize this scenario is via a game tree such as that depicted in Figure 2.

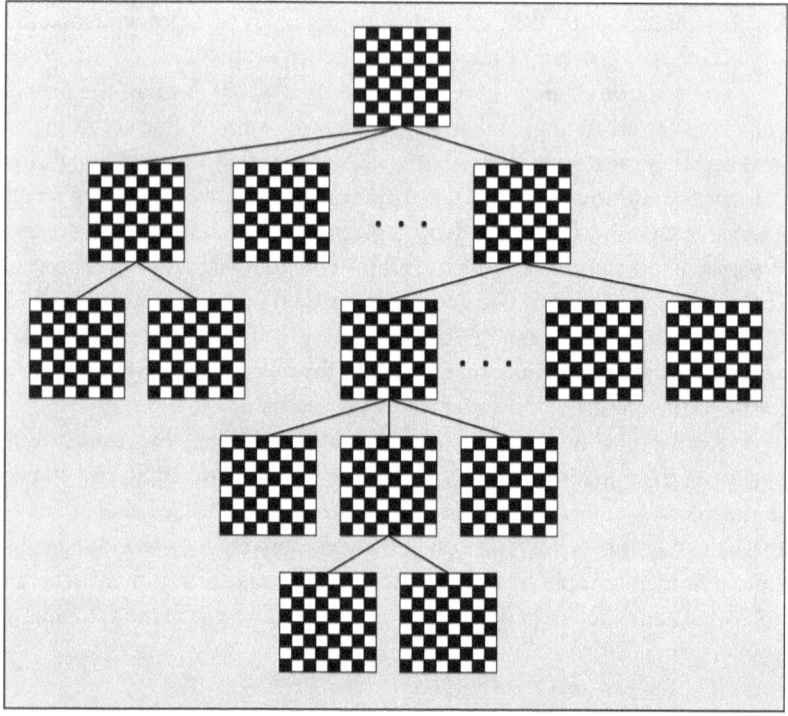

Figure 2. A partial game tree for chess or checkers

Assume that the board at the top of the diagram represents the state of the game board at the time of your current move (imagine pieces in any configuration you wish). Each board at the next level corresponds to one of your allowable moves and is intended to show the state of the board after that move. For each of your moves, your opponent has a specific set of possible moves she can make, and so forth. If we follow any particular path from the top node down into the tree, we will arrive at a representation of the board showing the results of executing the moves along that path. The tree is merely conceptual, so we can evaluate possible results arising from various move sequences and use that to guide our decision about which move to actually make. Down one path we come to a board where we find ourselves in dire straits (e.g., perhaps we have lost a queen and other key pieces); down another we find the piece count to our advantage. Down some paths lies certain victory; down others, certain defeat.

Humans, however, can't look very far ahead, if for no other reason than the memory constraints described earlier in this chapter. Consequently, it should be very apparent that any choice of move is just an approximation of sorts. Computers, on the other hand, are able to faithfully recall many more board positions and one might imagine that they could look far deeper into the tree. That is correct and is what most chess and checkers playing programs do. However, despite the fact that the best human chess player was defeated by a computer years ago, it would be a mistake to think the computer is immune from the need to make a best guess. The measure of complexity for this type of problem is the reason. You probably noticed that Figure 2 only depicts a partial game tree indicating a maximum of four moves (two for each player) and not even showing all possible moves at each level. The reason is simple—as one moves down the tree, the number of boards that must be represented literally explodes. If for each board we assume only ten possible moves (which is quite small for chess early in the game), the number of boards goes from ten at the second level to one hundred at the third, 1000 at the fourth, and so forth. This exponential growth means that there are a billion possible paths to level ten and a trillion to level thirteen.

In case the significance of this hasn't dawned on you, think of it this way. In 1950 Claude Shannon of Bell Labs estimated that it would take a computer checking one million boards per second more than 10^{90} years to search the complete game tree for chess.[27] That's a mite longer than most people are willing to wait. Deep Blue, the IBM computer that defeated reigning human chess champion Gary Kasparov in the 1990s operated 200

---

27. Shannon, "Programming a Computer." If you're notation-challenged, that's a 1 followed by 90 zeros.

times faster than Shannon's estimate[28] but reducing the time to 10^{88} years is not much of an improvement. The bottom line is that, unless the number of remaining pieces is quite small, even the world's fastest computer must stop somewhere long before it comes anywhere close to the leaves of the tree and then approximate how good the boards at that level appear to be. There is no chance for optimality, just an informed guess.

Because this section was heralded as being about significant issues, it is possible that you are wondering why I would spend much time discussing a game. I have done so because this example is typical of a broad class of problems whose very nature (i.e., their complexity category) limits their solutions to approximations. We call such problems intractable and, except for very simple cases, we can never know for sure if the answers we obtain for them are the best ones. As computers get faster, truly intractable problems will remain so even though we can solve more and more of them in a given amount of time. For example, a chess playing computer that was a million times faster than Deep Blue would still take millions of billions of years to search the entire game tree. It could, however, search deeper into the tree, meaning it would possess the potential for better approximations.

If human brains are getting faster, however, the time frame must be measured in millennia, not years. Although we can rely on technology to speed some of our searches, it can only take us so far. A computerized dating service might be useful for identifying promising leads but most of us are probably not yet willing to make the final selection of a mate based on advice from a machine. Furthermore, even though a typical computer search for a date is not intractable, actual dating might as well be because it relies on much slower sequential physical encounters.[29]

But dating is just one of many significant things that are characterized by our inability to exhaustively explore the relevant search spaces before we make a choice. I'm thinking of things such as where to go to college, which major to choose, what occupation to pursue, and the like.[30] For such options, approximation is the rule and your faith that a particular set of alternatives is in your best interest may be well rewarded or not. Somewhere in imaginary parallel universes a number of your clones who believed they

---

28. Cf. Krauthammer and Dowell, "Kasparov."

29. Although it may take nanoseconds to evaluate a prospect in a database, one needs at least a couple of hours even for a short date and much longer for a more thorough evaluation!

30. Selecting a university may provide some limits on whom to date but there will always be vast portions of the space we can never know anything about that could contain a better answer than the one for which we settle.

were best served by different choices may be happier, wealthier, and wiser.[31] Hopefully, most are not.

Certain Types of Measurements Are Always Approximations

When my younger son was an infant we had occasion to visit his pediatrician twice within a two-week period and, as was the custom, the nurse recorded his height and weight each time. You can imagine my surprise to discover that he had grown an inch and a half between the two trips! As anyone with a calculator could see, at that rate he would outgrow his bed (not to mention his bedroom) before he reached six and that by the time he entered college he would be approaching sixty feet in height and ready for his own sci-fi movie.[32] Though I'm saddened at the loss of potential income, I am happy to report that he topped out at a shade less than six feet and began university life still able to wear hand-me-downs.

It was clear way back then, however, that the issue was not his phenomenal growth or even my mathematical skills. At least one of the two measurements was almost certainly in error. However, if either was precise it was a matter of pure luck and not just because it is impossible to completely uncoil a baby. The truth is, certain types of measurements must always be approximations. If you doubt this, try to explain what would constitute an exact measurement for someone's height. Supposedly, I'm six feet tall barefoot. But according to whose ruler? Is that before or after taking a muscle relaxer? Shouldn't the temperature matter? Does the measurer have astigmatism?

Long before my children were born, I had already begun to suspect that there might be something fishy about the act of measurement. My epiphany occurred in freshman chemistry lab where we were periodically required to weigh various amounts of substances on ultra-sensitive scales. I am sure those instruments were expensive and had been purchased with the express intent of improving the ease and accuracy of our experiments but, try as I might, I could seldom get a reading about which I felt confident. Merely placing the sample on the scales would send them into a long-lasting oscillatory mode that was perpetuated by the slightest vibration, perhaps due to someone jumping on the other side of campus or the breath

---

31. I allude to Everett's many-worlds hypothesis (cf. Byrne, "Many Worlds") only as a way to illustrate the options for an unknown future and not to claim that those parallel worlds actually exist.

32. This kind of extrapolation is reminiscent of Mark Twain's whimsical look at the actual shortening of the Mississippi River over time (*Life on the Mississippi*, 136).

exhaled in a curse carelessly directed toward the device itself. Eventually, I would make my best guess and move on but somehow my approximations were never as good as those the professor was anticipating. But I was only a freshman. How could I know that even the professor himself could do no better than an approximation?

Doubts about measurement followed me into my physics classes as I tried to understand how anyone could make bold statements about invisible entities based on what they read on some gauge.[33] As I eventually learned in quantum theory, there were good reasons for my concerns (although it was too late to help my chemistry grade). The sovereign controller of measurement accuracy at the subatomic level is called the Heisenberg Uncertainty Principle. If we are attempting to measure the position and velocity of an electron, for instance, this informs us that the accuracy with which we desire to know one of these quantities limits the accuracy with which we can know the other. The physicist accepts this news with about as much enthusiasm as he would a gift of health and insight subject to the caveat that the more he had of one the less he could possess of the other. Be that as it may, let's assume that we select one of the variables (position or velocity) and determine to make it as exact as possible. Unfortunately, we are faced with the dilemma of being unable even to say precisely what we mean by "exact." All we can do is get below some agreed upon margin of error and call that "exact."

Thus our quantum quandary is more than a metaphor, although it is surely that. We face an intrinsic inability at all scales, not just the subatomic, to know when our measurements are exact because there are limits to our ability to define exactness. The same problem was encountered when we considered how to determine a person's height. Approximations are a natural result of the fact that measurement requires both a standard and a comparison against that standard. We'll consider each of these in turn.

Measurements Require a Standard

Occasionally I require students to empirically derive a value for the acceleration of an object falling under the influence of earth's gravity. This is a simple experiment but gravity, fortunately, is not constrained by their results or we would be in danger of floating off into space with the least provocation or being crushed under our own weight.

---

33. I doubt that this is what the writer to the Hebrews had in mind when he described faith as "evidence of things not seen" (Heb 11:1, King James Version) but the physics parallel is hard to miss.

Although the tools used by the students are not the sole determinant for the quality of the measurements, they play a significant role and this is true in any measurement. In fact, having reliable standards is deemed so important that there is an international organization, the Bureau International des Poids et Mesures in France,[34] whose sole purpose is to establish and guard Platonic ideals for length, mass, time, and so forth. For example, the Bureau defines the meter as "the length of the path travelled by light in vacuum during a time interval of 1/299,792,458 of a second."[35] That, of course, requires both a definition for "second" and a measurement of the speed of light. The Bureau obliges us regarding the first of these by defining a second as "the duration of 9,192,631,770 periods of the radiation corresponding to the transition between the two hyperfine levels of the ground state of the cesium 133 atom . . . at rest at a temperature of 0 K."[36]

If you have to ask what that means, it means you can't make the measurement and must trust someone else to do so! Presumably, great care is taken by the employees of an international organization with a French name but your trust rides on the hope that poor vision, hangovers, or the transposition of a couple of digits do not contribute to the published value.

What you should notice is that approximation is inherent in even the most carefully established standards themselves. Although someone can decide absolutely that this or that number will be the basis for a given standard (i.e., as for the meter or second), actually obtaining any such values requires the use of a machine capable of counting into the millions or billions at a phenomenally rapid rate and without error. If Plato had left us such devices we might feel more secure but each is produced by engineers subject to all the foibles that plague the persons who use them.

Furthermore, the quixotic quest for Platonic standards has led to changes in definitions, meaning individuals in the past had inferior bases for their measurements and suggesting that ours may not be the final answer. It is also possible that, even with a constant definition, the standard itself can vary. The standard for the unit of mass (the kilogram), for instance, is a physical object subject to change due to the collection of molecules in the atmosphere. Part of the kilogram's very definition, therefore, deals with the need to periodically clean that piece of metal.[37] You might be inclined

---

34. The International Bureau of Weights and Measures (www.bipm.org/en/bipm/).

35. http://www.bipm.org/en/si/base_units/metre.html.

36. http://www.bipm.org/en/si/base_units/second.html. Note that the impossibility of reaching absolute zero or making any measurements at that temperature even if it could be reached have not prevented the standards makers from incorporating it into their definition for the "second."

37. Cf. Girard, "Third Periodic Verification."

to protest that my nitpicking involves issues which are too small to worry about—fluctuation in the mass of the kilogram, for example, is a matter of millionths of a gram—but the Bureau does worry about them. Assertions that all we really need are reasonably good standards are acceptable only if we can agree on what is reasonable. Presumably someone's foot was good enough at one time as a standard for length but even a precisely machined platinum-iridium bar (the old standard) is no longer adequate. Consequently, when it comes to what we are willing to tolerate, "small" is relative and an implicit acknowledgment that even standards are approximations.[38]

Measurements Require a Comparison Against a Standard

There is no such thing as an absolute measurement. Every measurement is relative to some standard and the act of comparison is frequently more problematic than the standard itself. So as not to overstate the case, let me make it clear that some comparisons appear to contain no possibility for error. If Bob is Sam's month-old son, it seems plain that we could make an error-free claim about who is taller. This is the nature of many discrete measurements when the number of entities is few and the differences are (apparently) great. However, if we want to know how much taller Sam is than Bob, we have a different dilemma. In the former case either Sam or Bob could serve as the standard but now we are forced to employ another one against which we can compare the duo.

Lest you think that the problem arises only because of difficulties associated with assessing the length of a person, imagine comparing rigid items with well-defined ends—say a "meter" stick and a "twelve inch" ruler. Next, assume that we could get our hands on the Bureau's ultrafine, well-calibrated instruments and that we were provided with extensive training in their use. What then? Because the task is not just to say that one item is longer than the other but by how much, any answer will be an approximation (despite our willingness to accept incredibly small margins of error).[39] Even the Bureau of Standards couldn't have helped Shylock safely obtain his pound of flesh.[40]

---

38. For example, the time of day displayed on the NIST website (www.time.gov) is good enough for most purposes but cannot be exact (as noted on the page itself) because of communications delays (among other things).

39. Showing measured values with "+/-" symbols (e.g., 6.27 +/- 0.02) or including error bars on charts are typical ways to designate the approximate nature of data but their absence should not be construed to indicate exactness.

40. Shakespeare, *Merchant of Venice*.

* * *

In my state, drivers are required to renew their licenses every four years. At my last renewal I happened to notice that the entry for hair color was "gray." Curious, I looked at the older license and it, too, showed "gray" for the color. There was a part of me that was just grateful to have enough hair to show but I was left wondering, "When did that happen?" At some time in the past it was deemed brown by the license mavens. How many hairs had to change color and by how much? When did it become "gray"?

I ask this because it might be easy to get the idea from the examples I have employed above that measurement is primarily of interest to scientists, engineers, and physicians. But, it is hardly as esoteric as that. Those illustrations were chosen to make it clear that even the most precise measurement boils down to a statement about what is believed to be the case. In the world of everyday experience the role of approximation is much easier to see, even if the approximations themselves are not always easy to make or the consequences easy to take.

Judging speed with a radar gun, assessing voices for a choir, or performing a survey are all measurement activities where even slight variations can have significant repercussions. Neither does it help that the environments where such measurements occur are often messy or that the measurements themselves may be meaningless under different circumstances. (Was his time for the forty yard dash before or after being attacked by a 350-pound lineman?)

In fact, the greatest ambiguity may occur with the most important measurements. The biblical book of Daniel records that the Babylonian king Belshazzar was "weighed on the scales and found wanting,"[41] apparently because he was a featherweight in the realm of humility and piety, but one can imagine other ways to assess human value. Martin Luther King Jr., for instance, noted that, "The ultimate measure of a man is not where he stands in moments of comfort and convenience but where he stands at times of challenge and controversy."[42] That sounds great but compare it with an alternative view that, "The true measure of a man is what he would do if he knew he would never be caught."[43]

Just deciding which standard to apply, then, can be a tricky business (and entails its own type of measurement) but until we have taken that step

---

41. Dan 5:27.
42. King, *Strength to Love*, 26.
43. Attributed to Lord Kelvin.

we can't hope to "measure a man." The range of public opinion related to Presidential performance, for example, is due in large part to the application of different standards to data that are often undisputed. Even when we are able to agree on a standard, all sorts of factors will usually make it difficult to reach a consensus as to how the ratings should go. One result is that we end up expressing many measurements as averages—not because we are lazy or incompetent but because it is the only approach open to us.[44] Pollsters also know that just making some measurements can affect the results but this is not news to quantum physicists, anthropologists, or parents.

In short, measurement is more than my personal nemesis. Many of the things for which we would like to obtain an accurate assessment are forever constrained to approximations and we must be satisfied with trying harder to enhance the precision of the beliefs those approximations represent.

Theories Are Just Approximations

Scientific theories are the offspring of a fruitful union between conjecture and experimentation but such mergers do not happen lightly or quickly. At any point during the affair the results of some experiment can be just as catastrophic as a bad date and the relationship can shatter. Over time, however, provided no such irreconcilable differences are detected, the union can be consummated and the conjecture (i.e., hypothesis) elevated to the status of theory. But all experiments involve measurement and so, as a consequence of the limitations discussed above, there is a very basic sense in which we can say that theories are just approximations.

I could let it drop there but a common tendency toward unwarranted exaltation of scientific theory suggests that its approximate nature is underappreciated if not ignored outright. People with little background in science or mathematics, for example, are sometimes prone to confuse theorem and theory but the former is a mathematical term for which claims of exactness and factuality can be legitimate (i.e., assuming the application of logically sound reasoning to a valid set of starting premises). Yet no such claims can be made for even the best scientific theories, if for no other reason than it is impossible to say which is the best. All we can hope to say is what is best today and even that may be contentious.

Unfortunately, it is possible even for people who know a great deal about science to make gratuitous assertions about theories, particularly ones not universally accepted. Thus we find individuals such as Carl Sagan bluntly

---

44. Or, the "measurement" might be the result of an interpolation or extrapolation, as with speedometers, cyclometers, and heart-rate monitors.

announcing that, "Evolution is a fact, not a theory."[45] Richard Dawkins is a bit more grandiloquent than Sagan, proclaiming that

> One thing all real scientists agree upon is the fact of evolution itself. It is a fact that we are cousins of gorillas, kangaroos, starfish, and bacteria. Evolution is as much a fact as the heat of the sun. It is not a theory, and for pity's sake, let's stop confusing the philosophically naïve by calling it so. Evolution is a fact.[46]

Lest you think I am about to tie God's hands by making this a religious issue, allow me to share the perspective of the Jesuit paleontologist Pierre Teilhard de Chardin, written over twenty years before Sagan penned his words.

> We still find here and there in the world people whose minds are suspicious and skeptical as regards evolution . . . And because biologists continue to discuss the mechanisms by which species could have been formed, they imagine that biologists hesitate (or that they could hesitate without suicide) about the *fact* and reality of such a development.[47]

You should note that none of the scientists I have quoted (an agnostic, an atheist, and a Christian) are talking about some watered-down version of evolution.[48] Furthermore, I want to make it plain that I have no intention here to castigate evolution. It is far and away the best (and arguably the only) scientific theory we have for the biological progression of life. However, if one wishes to call it a fact he must accept that he is moving it outside the realm of science as we know it. That is a step I doubt any of these would want to take.

The philosopher Philip Kitcher, in an attempt to defend the evolution-as-fact perspective, suggests thinking of "fact" as, "something so amply confirmed by the evidence that it may be accepted without debate."[49] Evolution, he maintains, can make that claim. But what part of evolution does he mean? The basic theory is an insightful observation but the concept of natural selection is too broad to provide more than a general framework from which to proceed. Even Dawkins, in the same article in which he calls

---

45. Sagan, *Cosmos*, 27.
46. Dawkins, "Illusion of Design."
47. Teilhard de Chardin, *Phenomenon of Man*, 138, emphasis mine.
48. The primary difference is that Teilhard de Chardin thinks God could have done it no other way while Sagan and Dawkins would replace "God" with "nature."
49. Kitcher, "Fact of Evolution." Compare Locke, *Human Understanding*, book IV, chapter XV, #2.

evolution a fact, acknowledges that, "Darwinism proves to be a flourishing population of theories, itself undergoing rapid evolutionary change."[50] I suppose one could try to distinguish between evolution and Darwinism—many have done so—but the devil is in the details.[51] This is why, as Teilhard de Chardin noted, "biologists continue to discuss the mechanisms."[52] Perhaps it is also the reason that paleontologist and evolutionary theorist George McGhee believes, "the modern scientific discipline of evolutionary biology is in a similar position as the scientific discipline of chemistry before the discovery of the periodic table of elements."[53]

The value of most scientific theories is gauged on their predictive power but, in the absence of the more detailed mechanistic theories to which Teilhard de Chardin and (I believe) Dawkins refer, the predictions are quite vague. Although over time many of the details will likely be fleshed out (as many already have been), unless evolution is somehow privileged beyond other scientific theories, these efforts will not be without significant scientific give and take and, in many cases, not a little controversy.[54]

By now, you may have lost sight of the reason we went down this path in the first place. It was necessary, however, to present a fairly involved discussion of evolution as theory because it provides a current and much-publicized example from which to illustrate the approximate nature of scientific theory in general—even those theories that are sometimes construed as impregnable.

The history of science, of course, is replete with examples of the changing nature of revered theories. From Aristotle's concentric spheres to Ptolemy's epicycles, men struggled to keep the earth at the center of things. However, as we laugh at the naïveté of such views, we may forget that those theories were accompanied by long and serious observations made by humans arguably just as intelligent as Newton or Einstein. Alexander Pope's catchy characterization of Newton—"Nature and nature's laws lay hid in night; God said 'Let Newton be!' and all was light."[55]—suggests a potentially different fate for

50. Dawkins, "Illusion of Design."

51. Most biologists call the evolution of species a "fact" in a sense similar to which most Christians call the resurrection a fact—both beliefs are based on what their proponents consider reasonable evidence.

52. Teilhard de Chardin, *Phenomenon of Man*, 138.

53. McGhee, "Convergent Evolution," in Morris, *Deep Structure of Biology*, 17.

54. For one thing, actual details about past events must always be inferred, even with a preponderance of evidence arguing for some interpretation. In addition, even if a theory appears to be completely accurate in predictive power, a simpler one might be just as accurate.

55. Attributed to Alexander Pope.

Newton's insights, but his laws of motion succumbed to change (due to relativity theory) far more quickly than Aristotle's take on the universe. Einstein's theories have, likewise, already been subject to modification.[56]

It is especially interesting when something hitherto declared scientifically impossible (under one theory) is shown to be possible (potentially leading to a theory change). Russian chemist Boris Belousov's paper documenting a novel chemical process was rejected for publication in the middle of the twentieth century because the idea was deemed ridiculous. Several years later another Russian scientist duplicated Belousov's feat but also managed to get someone else to take notice, and the Belousov-Zhabotinsky reaction is now a classic example of non-equilibrium thermodynamics.[57] Such processes, however, were beyond the grasp of those locked in the old mindset.

This grip of scientific belief occurs within what philosopher of science Thomas Kuhn termed a paradigm.[58] For Kuhn, the practice of "normal science" transpires within a prevailing theoretical system (i.e., paradigm) viewed more or less as truth by those so engaged. Weaknesses in the theoretical underpinnings (perhaps revealed in some experiment) can lead to overthrow of the current paradigm but it can be painful and resisted.[59]

Thus, although science is rightly thought of as a scrupulous method for pursuing knowledge, understanding, and even truth, Kuhn's insights remind us of the tenuous nature of even the most rigorously tested theories. The "truth" of one generation of scientists is the approximation of the next and we need have no illusions about "exactness" in science. Every theory can be placed on a scale of belief and, with time, they will occupy various positions along that scale.

But this is not just true of scientific theories. Despite the fact that I have focused on that domain in this section, science doesn't have exclusive rights to the term "theory." There are theories of history, theories of religion, and popular theories on just about anything you can imagine. Perhaps some of these are poor uses of the term—we don't always use our language with integrity—but all are approximations and we should be concerned about

---

56. The best known is probably his "cosmological constant" in the theory of general relativity (cf. Isaacson, *Einstein*).

57. Strogatz, *Sync*.

58. Kuhn, "Essential Tension" and *Structure*.

59. Ibid. Consider the role ascribed by Kuhn to faith: "The person who embraces a new paradigm at an early stage must often do so in defiance of the evidence provided by the problem solving. He must, that is, have *faith* that the new paradigm will succeed with the many large problems that confront it, knowing only that the older paradigm has failed in a few. A decision of that kind can only be made on *faith*" (Kuhn, *Structure*, emphasis mine).

any claims for a complete theory for quantum mechanics, evolution, causes for social conflict, or God.[60]

Even the Most Precise Formalisms Have Built-in Limitations

In the previous section I made it a point to contrast theories and theorems and went so far as to claim that theorems could be viewed as truth if they were built on valid premises and employed sound logic. What I neglected to mention is the problem we have determining what constitutes valid premises.

The German mathematician Kurt Gödel upset a number of mathematical apple carts in 1931 with his proof that within any mathematical framework there will always be propositions that cannot be proven.[61] You may have suspected this in some mathematics class where you were given certain axioms to use as the basis for all sorts of proofs but were never asked to prove the axioms themselves.

Gödel's theorem or its spinoffs have been employed to support a variety of views, not all of them compatible. Penrose, for instance, uses it to reason that human thought cannot be captured by a computer.[62] Others, who accept the algorithmic nature of human thinking, simply see Gödel's theorem as an indication that there will always be things the human brain can't do.[63]

Regardless of what you think about such things (or whether you have thought about them at all), it is not necessary to fully understand everything implied by Gödel's theorem to appreciate that this important mathematical result has implications for faith. If there are propositions that neither you, I, nor anyone else can prove—not because of our mathematical incompetency but because of the very nature of mathematics itself—then we can only accept them as true in the belief that they are so. It's not that we have any particular reason to doubt those axioms but Gödel showed us that we have every reason to doubt the possibility of establishing their credibility

---

60. God may be immutable but our theories about him are not. Paradigm shifts in religion (cf. Barbour, *Religion and Science*) are no less momentous than in science and can be seen in such areas as changing views on law and love, works and faith, sacrifice, etc.

61. Gödel, "Über Formal Unentscheidbare." Hofstadter describes Gödel's Incompleteness Theorem thus: "All consistent axiomatic formulations of number theory include undecidable propositions" (*Gödel, Escher, Bach*, 17).

62. Penrose, "Can a Computer Understand?" in Rose, *Brains to Consciousness*.

63. Cf. Turing, "Computing Machinery" and Moravec, *Robot*.

on mathematical grounds. When we discover boundaries on what we can honestly pronounce as fact even in the most precise and trustworthy of disciplines, we might suppose that less structured human endeavors are even more limited.

The More You Know, the More I Must Accept on Faith

I frequently return from visits to my university library soundly humbled. I go with the express idea of increasing my knowledge and I am usually successful. But I return to my office knowing that the insights which I have acquired in the course of that visit, even those I have managed to accumulate through years of study, pale in the light of countless treatises in my own specialties and look downright anemic when compared with the vastness and vigor of human knowledge in the realms beyond. For every book I pull from its shelf there are thousands that remain untouched. From all I can tell, it will only get worse. It has now become all but impossible to stay abreast of even a single significant discipline, for as one ascends the tower of knowledge he finds that the edifice itself is rising faster than he can climb.[64]

More Knowledge Means More Faith

The good news is that human knowledge is undergoing its own big bang kind of expansion. The bad news is that you and I occupy small outposts on the fringes of all that knowledge and the fraction thereof which any one individual can conceivably know is becoming smaller at light speed. The paradoxical result is that the increase in knowledge, far from removing the need for faith, actually increases it.

My doctoral dissertation, for example, involved creation of a system to simulate aspects of human intelligence on a computer. During the defense I explained my model to members of the committee (and others who couldn't find something better to do with their time), employing numerous equations, diagrams, tables, and examples to illustrate how the system was constructed and functioned. Slides proclaimed that the system had the ability to perform tasks relating to mathematics, reading, music, and memory strategies (among

---

64. Arthur Peacocke says it this way: "If the world were a closed system, we would expect an ultimate convergence in our knowledge as it accumulates, but nothing like this seems to be happening. Our awareness of our ignorance grows in parallel with, indeed faster than, the growth in our knowledge" (Peacocke, *Intimations of Reality*, 58).

others). Because I passed that exam, I suppose the committee believed me, but not a single member had actually seen the model in action.

If you have little experience in science, this may seem bizarre but it is the rule, not the exception. In principle, my committee members could have run the experiments described in my presentation and thesis, even though they did not. It is this same "in principle" concept on which much of science operates. I had not attempted to replicate the work of the many sources I cited in my dissertation, nor could I have. This state of affairs is repeated thousands of times each day as scientists read and (usually) take the word of someone else regarding the results of their efforts.[65]

Occasionally there are doubts about the claims (as with Belousov's work described earlier) with no effort at replication. Sometimes doubts, curiosity, or a desire to expand an experiment lead to genuine attempts to repeat it. Once in a while those attempts fail, as occurred with cold fusion late in the twentieth century.[66] By and large, however, once an article makes it to print its conclusions are taken on faith. This doesn't mean that each reader accepts it equally—the responses will always occur on a continuum of belief—but in many (perhaps most) cases there will be no first-hand knowledge (i.e., primary observation) of the claims made outside that of the originator.[67]

The way in which one knows how much faith to put in something is primarily a matter of experience but, as Kuhn reminds us,[68] experience can sometimes be blinding rather than eye-opening. In general, readers expect the assertions to appear plausible and to offer good prospects for replication. Perhaps they know and trust the author. It should be clear, however, that this is necessarily a subjective evaluation built on each individual's unique background.

What I have just described is characteristic of every field of science but it is not limited to science. When we read a biography, we trust the historian. When a disk jockey announces that a piece of music was performed by this or that musician, we take him at his word. When we view a painting at the

65. This occurs during article review and, if the paper is accepted, after publication. Although many things can prevent an article from receiving a positive review, significant doubt about the author's claims is a killer. One thing is certain, anyone reading a published article is not only placing faith in the author but in the reviewers as well.

66. Browne, "Physicists Debunk Claims."

67. However, errors in an experiment that is never repeated exactly can become obvious in related experiments. Furthermore, with many people engaged in the same type of research there is actually more integrity overall than when there were only a few. Nevertheless, the knowledge edifice on which any scientific discipline is built remains impossible to personally verify.

68. Kuhn, *Structure*.

museum, we believe the inscription on the plaque that identifies the artist.[69] In principle we could run down the facts ourselves but in practice it is impossible for all but a limited number of cases.

Gaining the competence to evaluate something thoroughly is a time-consuming task. I spent eight years in graduate school learning how much I didn't know. When done, I like to think that I was relatively competent but as one of my professors reminded me, it was only in one micro-domain of knowledge. As the French philosopher Maritain noted, "no man can . . . specialize in all branches of science, a contradiction in terms. He is fortunate, indeed, if he can make himself master of a single science."[70] Or, as science writer John Horgan describes the linguist Noam Chomsky's view, "Modern science has stretched the cognitive capacity of modern humans to the breaking point. . . . In the nineteenth century, any well-educated person could grasp contemporary physics, but in the twentieth century 'you've got to be some kind of freak.'"[71]

There is no need to restrict these views to science but let's suppose that such a freak begins graduate work immediately after finishing college and is able to complete the study and research necessary for a Ph.D. in five years. Furthermore, imagine that this student is not content to stop there but pursues degree after degree, determined not to be confined to one small area of expertise. If the five years-per-degree pace can be maintained, by the time this person is in her early seventies she will be an expert in ten micro-domains of knowledge.[72] Compared to other people, this hypothetical person would be considered extraordinarily intellectual but that's a poor standard. Relative to the ocean of knowledge, drinking with ten straws rather than one won't make much difference.

---

69. Cf. Lewis, *Mere Christianity*, 53–54.

70. Maritain, *Introduction to Philosophy*, 84. Physicist Erin Schrödinger echoes Maritain's sentiments: "We feel clearly that we are only now beginning to acquire reliable material for welding together the sum total of all that is known into a whole; but, on the other hand, it has become next to impossible for a single mind fully to command more than a small specialized portion of it" (Schrödinger, *What is Life?*, 1). The task is much more daunting now than when Schrödinger penned these words in 1944.

71. Horgan, *End of Science*, 153.

72. Actually, this is a generous overestimate of the student's competence. While pursuing one degree much of what was learned for another could become outdated or forgotten.

God of the Gaps

We can get an even better handle on the role faith plays in a world of expanding knowledge by considering the so-called "God of the gaps"[73] perspective. This is the idea that, as human understanding increases, the need to postulate God as an explanation decreases for things once considered incomprehensible. Some would go as far as to suggest that humans invented gods to fill their need for an explanation of things they couldn't yet understand but that, as human knowledge grows, the need for God diminishes proportionally. Figure 3 suggests how, according to this perspective, gaps in understanding are plugged over time by growing insight made possible by various intellectual endeavors.

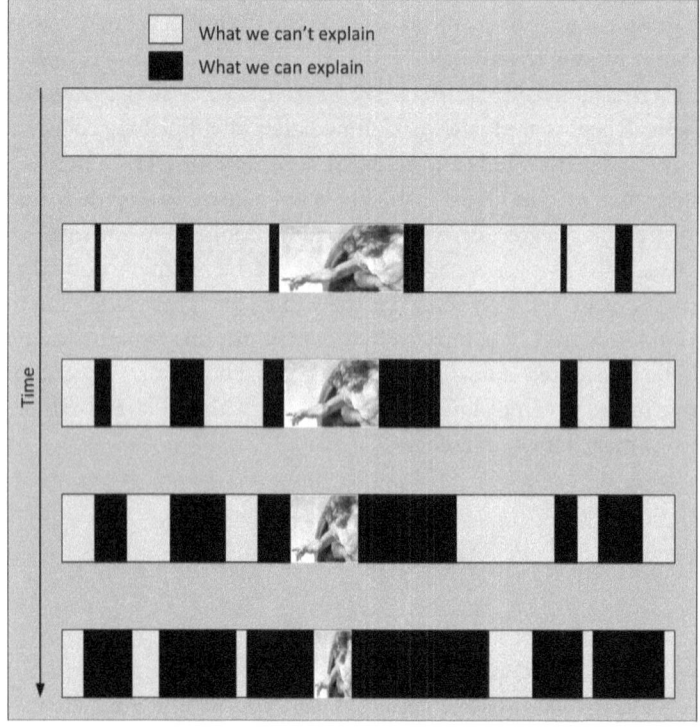

Figure 3. In a "God of the gaps" scenario, growing human understanding is supposed to diminish the need to use God as an explanatory construct. The diagram represents several snapshots in time.

This point of view is not without merit because it helps focus attention on the benefits of trying to understand how things work as opposed

---

73. Cf. Drummond, *Ascent of Man*, chapter X (originally published in 1904).

to merely being content with the fact that they do.[74] However, "God of the gaps" is far too simple to provide much help beyond this rudimentary insight and it is easy to read into the concept more than is warranted.

One of those erroneous inferences is to treat this concept as "faith of the gaps"—that is, that we needed faith once upon a time pertaining to all sorts of things but that human intellectual progress has diminished that need. But even though an increase in understanding can change one's faith (perhaps turning it in a new direction), the idea that becoming more knowledgeable somehow removes the need for faith is unsupportable. There are, however, some additional problems with the "God of the gaps" (or "faith of the gaps") outlook.

In the first place, the explosion of human understanding that is supposed to shrink the gaps can generate precisely the opposite effect. We have already noted that the more we learn, the more we discover that there is so much more to learn. The "God of the gaps" viewpoint, however, suggests a static universe of potential knowledge and discernment. Figure 3, therefore, is too simple, and there is a vast expanse of things unknown both to the left and right of any of the time periods represented in that finite diagram.

The dynamic nature of understanding also makes it plain that the gaps themselves are not fixed. What once appeared to be a small crack in our comprehension may suddenly loom as a yawning crevasse or even a whole host of them. Seeking to close one gap, we often open several others and this multiplication occurs specifically because of an increase in our insight. Fauconnier's and Turner's description of this dilemma for cognitive neuroscience—"Phenomena that were not even perceived as problems all have come to be regarded as central, extremely difficult questions"[75]—is applicable to many domains of inquiry. As Haught puts it, "the deeper science digs, the more impressive is the extent of the mystery it uncovers."[76] Consequently, any proposal to the effect that closing gaps will eventually cause them all to disappear is unsupportable in the face of how things really work.

Furthermore, we have been assuming all along that we are able to close (or at least bridge) some gap as our ability to explain grows. But what does it mean to "explain" something? Many scientific "explanations" take the form of equations—the math is the explanation—but what kind of explanation is

---

74. Part of the consternation over the perceived substitution of God for explanation in the "God of the gaps" view was that there was really no explanation in that approach—just a proclamation regarding agency. We still want to know how he did it (cf. Kaufman, *Jesus and Creativity*).

75. Fauconnier and Turner, *Way We Think*, 7.

76. Haught, *Christianity and Science*, 31.

that? Typically, it is a predictive kind, fantastic in many respects but sometimes leaving one a bit cold.[77]

Isaac Newton, for instance, showed that the gravitational force between two objects is proportional to the product of their masses divided by the square of the distance between them. This is a wonderfully satisfying and useful predictive model (i.e., it describes what gravity does) but, when attempting to explain what gravity is, Newton was at a loss.[78] Although it is obvious that he could describe it as an attraction between two bodies, that doesn't tell us much. For example, depending upon the bodies of which we speak, there are all sorts of attractive possibilities in addition to gravity—electrostatic, common interests, love, and so forth—for which we might seek an explanation.

Einstein's theory of general relativity refined the predictive power of Newton's results but, beyond that, we might be hard-pressed to decide whether an unexplained curvature in space[79] is to be preferred over an unexplained attraction. More recent theory indicates that, "bodies interact . . . by exchanging tiny subatomic particles,"[80] but why in the world would they do that?[81]

George Mallory was a mountaineer famous for his early attempts on Mount Everest but is probably best remembered today for his, "Because it is there," response when asked why he climbed. Although his answer might have some predictive value—i.e., you might expect that if he was anywhere near a mountain he would climb it—I imagine that one reason his remark has been remembered at all is not because he offered any satisfying explanation but primarily because it leaves us wondering, "What kind of explanation is that?" We would probably feel more comfortable (in the sense that we felt we had a better understanding of his reasons) if he had replied that he climbed because of the challenge, excitement, unknown, camaraderie, exhilaration of the heights, and so forth, but would we? Let's pick one of these alternative "explanations." Suppose he had said he climbed for the

---

77. Given A and A→B we can infer B (where A and B are statements about the universe) but that does not necessarily explain B. Also, see Lewis, *Discarded Image*.

78. Spielberg and Anderson, *Seven Ideas*.

79. Unexplained because, even though the theory "explains" that space-time is warped by the mass of an object, we are still left to wonder why.

80. Spielberg and Anderson, *Seven Ideas*, 102.

81. The "particle" is the graviton but its phantom nature—i.e., "generally believed to exist, even though it has not been observed experimentally" (ibid., 323)—currently requires significant faith. A common joke that parodies the abrogation of explanation for some hard-to-fathom concept does so with the phrase "and then a miracle occurs." I am not suggesting that we call gravity a miracle but, given what we have considered about efforts to explain it, we might be tempted to say, "and then gravity occurs."

excitement. Wouldn't we then want to know why climbing excited him? Couldn't we press the issue to get inside his head until we had literally done so, seeking to find an explanation in the firing of this or that population of neurons or the quantity of some neurotransmitter?

None of this should surprise us, nor should it make us fail to appreciate the magnificence of predictive models or think that there aren't other kinds of explanation provided by scientific accounts or other kinds of explanation besides scientific ones.[82] It should, however, keep us mindful that "explanation" occurs at different levels and that each "explanation" begs for another one[83] (or several more).

Thus, although a cursory understanding of the "God of the gaps" perspective portrays the retreat of faith before the relentless onslaught of human knowledge, that cannot be the case. What we actually find is that our growing comprehension of how the universe works either increases the number of gaps, make us better appreciate their size, or raises significant questions about the very nature of explanation and understanding. In fact, even our attempts to close the gaps require faith that we can do so. Occasionally we succeed but at other times, in our efforts to fill what we perceive to be a ditch, we discover we are facing a Grand Canyon instead and watch (or at least our ancestors watch) as our attempts are swept into a pool of stagnant ideas and theories. Overall, our successes do grow but, because of the impossibility of staying abreast of all the details, so does the need for faith.

Summary

As we have seen, it is fairly easy to be beguiled by the myth of certainty. Yet when we pause to survey the scene, we discover that many of the things we claim to know for sure we actually do not and cannot. This is not to suggest that some things are inexplicable (although there probably are such things) but, to recast an old adage, you can know a few things about many things or many things about a few things but you can't know everything about everything. Whether our approach to knowledge is characterized by a deep one-dimensionality or a shallow multidimensional agenda, we can never have firsthand knowledge of most things. Thus, when science fiction writer Arthur C. Clarke noted, "Any sufficiently advanced technology is

---

82. Cf. Bunch, "Two Stories."

83. Spielberg's and Anderson's perspective pertaining to physics (*Seven Ideas*, 102)—"every answer to a question always seems to raise a new question"—is applicable to all domains of inquiry.

indistinguishable from magic,"[84] he implicitly acknowledged for all of us (technological savants included) the fundamental nature of our limitations, the probabilistic acceptance of things we take for granted, and the hopelessness of understanding everything.

Clarke's law also makes us wonder how a society could be labeled "advanced" if the majority of folks see its products as magic. It is comforting to believe that we are among those who have sound explanations for the "magic" of electromagnetic transmission (television, radio, and cell phones), quantum tunneling (transistors), gene expression, time dilation, brain function, and the like but why have we chosen to believe those explanations over others? Why do we deem one political system better than another? What makes us suppose this or that religion is closer to the truth or that all are the same? How we decide what to believe is the subject of the next chapter.

---

84. Clarke, *Profiles of the Future*.

———— CHAPTER 7 ————

Crafting Rational Faith

Do you wish us then to set down two forms of persuasion, one that provides belief without knowing, and one that provides knowledge?[1]

—SOCRATES

The Goal of Faith

A NUMBER OF YEARS ago I flew into Pittsburgh late at night, rented a car, and drove to a hotel for a short rest. As I left for work the next morning, I was plagued by a growing suspicion that something wasn't quite right. Finally succumbing to my concerns, I pulled over after a few blocks and discovered a receipt confirming that I was driving a car rented by someone else![2] Returning to the hotel, I found my car just where I had left it the previous evening and executed a quick swap before someone could spot my mistake and possibly conclude that I was a thief.

Now it is not unusual for folks occasionally to mistake someone else's car for their own and even to go so far as to try to unlock it but seldom does one successfully drive away in the other vehicle. In any event, this scenario suggests an analogy to help set the stage for our considerations of faith and reason. Let's call this sort of behavior "transportation faith" and let our confidence that some particular vehicle belongs to us and constitutes our

---

1. Plato, *Gorgias*, 36.
2. Although I believed that I was driving away in the car I had rented, the probability value of that belief was less than 1.0. In everyday language, I had doubts. But doubt can only arise in the context of faith. In fact, *doubt is faith—in the other side*.

personalized mode of transportation represent a more general belief that a certain course of action is the best way to move us from one position to another in some space of possibilities (perhaps involving relationships, education, career, health, skill acquisition, or whatever). You might, for example, engage in some relationship with the conviction that doing so will transport you from an emotional abyss to a euphoric high or spend countless hours practicing tennis thinking that your efforts will earn you a spot on the team.

This analogy makes it easy to identify four important variables associated with any and all applications of faith, typified by the types of examples just suggested. These variables are:

1. the objects of one's faith
2. the significance of those objects
3. the probabilities associated with faith in those objects
4. the accuracy of the probabilities

As apparently innocent mistakes in vehicle identification suggest, faith can be misplaced—that is, placed in the wrong objects—even when (as in my unusual case) most of the signs indicate it is correct. Divorce, disastrous career moves, significant time expenditures with no perceivable benefit, financial investment failures, and so forth are all testimony to this fact. Naturally, misplaced faith need not be in objects as momentous as these. Realizing that a class one thought would be easy is not, finding that a new recipe tastes terrible, learning that, although one is an excellent skier, snowboarding cannot be mastered overnight, or making any of a practically unlimited number of comparable discoveries indicate that our faith about even the most mundane of things can be wrong.

As we've already seen, faith in anything does not have to be absolute. Because faith is a matter of probabilities, even a misplaced belief can be partial. My own "transportation faith" described above was sufficient to make me drive away in the wrong car but not so great that I didn't eventually acknowledge its limitations and look for evidence to raise or lower its probability value. Upon discovering the other renter's receipt, one probability (that I was in the right car) plummeted while another (that my car was still at the hotel) skyrocketed. What I hope you see from this example and its analogous extensions to virtually all areas of life in which belief is operative—which means all areas of life—is that just thinking the probability is high does not and cannot guarantee the validity of a belief. Just because something strikes us as reasonable, it can still be wrong. If faith is to be sound, rational, coherent, and justifiable, the probabilities must be accurate.

They do not, however, have to be high. A justifiable probability for obtaining heads when flipping a balanced coin is 0.5 (i.e., a fifty-fifty chance) and there is no merit (except to a casino owner) in your assuming a higher (or lower) value. Ultimately, it is the accuracy of the probabilities associated with one's beliefs that makes them better or worse than another's (or that distinguishes them from one's own beliefs at other times).

In short, even though all faith is based on some reasons, the reasons are not always good and the strength of a belief is not enough. Given sufficiently strong beliefs, we will accept something, promote it, engage it, or do it even if it is wrong—merely believing that it is right will be adequate incentive. Sometimes those erroneous beliefs will have notable consequences (as with the thalidomide-induced deformities in newborns in the mid-1900s) while others are held with little noticeable effect for a lifetime or even across multiple generations and may be seen only in retrospect to have any real importance (e.g., belief in a geocentric universe). We have already noted, however, that we can seldom be certain about the ultimate importance of things—that assessment itself being an act of faith—so it seems appropriate to want to maximize the accuracy of our beliefs.[3]

I believe it is in this spirit that Barbara Taylor defines faith as, "a radical openness to the truth, whatever it may turn out to be."[4] Such a fundamental embrace of the search for truth, however, is less a definition of faith and more a description of the essential ingredient for achieving the proper probabilities for our beliefs; for making faith substantive. It is this openness to truth that can, in very real ways, set us free.[5]

Whether we want that freedom or not is an entirely different question. Most of us have to look no further than ourselves to find examples of people who take such comfort in one belief or another as to protect it at all costs. In those cases, the truth is the last thing we want.[6] In light of our usual desire to

---

3. Because our appraisal of the significance of a belief is itself a belief (i.e., a belief about a belief), that significance is, therefore, just another object of belief and all four of the variables discussed apply in that application of faith as well. (This also applies to our belief about the accuracy of the probability we assign to a belief.) You might detect a recursive nature to this scenario (i.e., we can have beliefs about beliefs about beliefs and so on), but we typically are not competent to process the probabilities or even identify relevance of these beliefs beyond the second level (i.e., belief about a belief) without significant effort and in many cases we don't even get that far.

4. Taylor, *Luminous Web*, 6.

5. Although Jesus' claim that "the truth will set you free" (John 8:32) was provided in a specific theological context, it is relevant to any domain in which we are subject to being enslaved by false beliefs.

6. Or as Jack Nicholson bluntly remarks to Tom Cruise in the movie *A Few Good Men*, "You can't handle the truth." This might extend to the truth about self,

appear rational, however, this is a most curious state of affairs and it is hard to see much long-term advantage in adopting or holding onto an unreasonable faith in anything.[7]

Consequently, this chapter is predicated on the idea that maximizing the accuracy of one's beliefs (i.e., increasing their truth content) is a worthwhile objective.[8] Believing that such a goal has value is, of course, itself an act of faith[9] but being correct about our beliefs seems preferable to being incorrect and this is how we will proceed. In the course of exploring this goal, we will find it helpful to contrast faith with wishful thinking, consider the extent to which the pursuit of rational faith is within our control, analyze the requirements for acquiring and maintaining rational faith, and identify various impediments to such a quest.

Wishful Thinking

It is common practice for many religious folks to ask a blessing prior to consuming their cholesterol-laden meals but one has to wonder how much faith is warranted in such a request. Considering what we now know about the content of our daily fare and the health consequences of consuming its ingredients, it might be a tad presumptuous to believe God would spare us from something we do to ourselves. Yet the idea that I can eat whatever I want without potentially harmful repercussions is really not faith at all, but wishful thinking. Unfortunately, faith and wishful thinking are easily

---

relationships, or other beliefs held dear.

7. Reason cannot exist without faith but faith can and often does exist without reason ("reason" in the sense of rationality, not unsubstantiated or irrational "reasons"). The quest for rationality seems to be behind the following remarks by psychiatrist Carl Jung: "I don't believe. I must have a reason for a certain hypothesis" ("Face to Face"). Jung's "reason," of course, is really just a way to change the probabilities so as to make a "belief" take on the persona of irrefutable knowledge. Even what Jung claims to "know" is still a belief.

8. I'm distilling thousands of years of epistemology into the simple equation *truth = what is correct* (i.e., what really happened; the right explanation; what can be verified) quite aware that all of these other terms require explication. But we must have some materials with which to work and this is a starting point.

9. Pirsig identifies this role of faith but takes a less positive view of rationality (*Zen and the Art*, 275–6). However, although exalting rationality may not be popular with everyone, it seems preferable to promoting irrationality (which apparently thrives without much conscious help on anyone's part). Even Pirsig's analysis of a "qualityless" world has a rational flavor to it.

confused. If we are interested in obtaining and maintaining sound faith, however, it is imperative that we distinguish between the two.[10]

"Wishful thinking" (or "hope") is what we want to happen. Our hopes are a reflection of how things would be if we were in charge. Consequently, we can hope that something is true or that some event will transpire without any significant belief that it actually will. We sometimes hear the claim that some people just believe what they want to believe but it is more accurate to say that those people hope what they want to believe. Many people, for example, play the lottery with high hopes that they will win, but the odds of actually doing so are the only sound measure of a legitimate faith in winning and that faith must necessarily be low. An accurate faith is hardly sufficient to motivate the gambler—only hope can do that.[11]

For instance, the only legitimate quantitative measure for faith in the outcome for the roll of a balanced die is 1/6 so one would need a motivating boost from hope if contemplating a bet. One part faith and five parts hope may be adequate inducement for a gambler but it is hard to call such an optimist rational. The appearance of improved odds is, for sure, just an illusion but it is the same type of illusion that is always produced when wishful thinking is conflated with faith.[12]

The effects of fusing (and thereby confusing) faith and hope are as far-reaching as the areas in which we do so, which is to say all areas. In the face of an 80 percent chance of precipitation we will sometimes schedule an outdoors event anyway in the hope that it won't rain on our party, even though the rational component of our decision only stands at 20 percent. We may excuse ourselves by questioning the forecast accuracy[13] but, unless we know the prediction history of the meteorologist, we are merely falling back on hope. The anticipated outcomes of asking for a date, taking an

10. *The Wit's Dictionary* (Bowles) defines faith as "Throwing your heart over the bar and letting your body follow." Although this characterizes the way many people view faith, it is really a description of wishful thinking (blind faith).

11. Many people who understand the probabilities but still gamble claim to be motivated by pleasure. Perhaps—there may be a number of arguable minute contributions to any motivation. But, if there was no hope of winning, any other factors would quickly evaporate (i.e., the pleasure is intimately connected to the hope). If you doubt this, contemplate the pleasure involved in putting a match to a twenty dollar bill (i.e., where there is no hope of a positive return).

12. The gambler may maintain that he has a high level of faith in his chances but, no matter what he calls it, anything he estimates beyond a one-in-six chance is wishful thinking (or blind, poor probability faith). Unlike contrived examples, however, one is seldom able to know the actual probability that should be associated with some belief.

13. The National Weather Service has discovered a way to make forecasts with complete certainty. Consider the following infallible snow forecast: "chance of accumulation less than one-half inch possible."

exam, or following a physician's prescription are all gauged on some mixture of faith and wishful thinking—what we can honestly expect to happen and what we hope will happen. Religious thinking can also have this character though it need not always do so.

Despite the fact that it is often difficult and perhaps even painful to identify the relative components of faith and hope in our perspectives on any particular subject, it is an important undertaking. Although there will inevitably be those who try to live in a dream-world, we are, presumably, more successful attempting to conduct our lives on the basis of accurate predictions about what we can expect than on what we simply hope will happen. I may hope that my salary doubles in the next year, that I garner a Nobel prize for my scientific research and a Templeton prize for my contributions to thinking on science and religion, that enrollment in my department quadruples with the next incoming class, that I live to be 500 years old in perfect health, and so forth but the only high probability event associated with any rigid plans I might make based on those hopes is that I will appear a fool to those who know me.

Correctly distinguishing between faith and hope thus enhances our prospects for leading rational, successful lives but it also enables us to identify areas in which we can attempt to change the probabilities associated with our existing faith. I'll have more to say about this shortly but the thing to note here is that hope is a powerful motivator. Recognizing that what we have been calling faith is really just wishful thinking can encourage and inspire our efforts to increase our faith. This is all part of the search for truth but any honest effort in this regard acknowledges that the faith in question might well decrease. The results, then, may be nothing like we had hoped, but that is sometimes to be expected when wishful thinking is the primary motivator.

Occasionally we are so hostile to this idea that we will go to great lengths to protect our hopes. Wishful thinking can cause people to try to convince themselves of a probability for something and to ignore the actual probabilities supported by evidence. But attempts at rationalization are frequently irrational and trying to disguise wishful thinking by masquerading it as belief is not to our credit. It is not unusual, for example, to find individuals living in the hope that they do not have some particular health problem (such as cancer), despite the fact that they possess a variety of symptoms and have managed to find creative ways to rationalize each of them. When the diagnosis finally comes, the difference in reasonable belief based on dispassionate evaluation of the evidence stands in stark contrast to what is then clearly seen to have been wishful thinking. In other cases it may prove equally easy to fabricate (rather than ignore) evidence to support some particular hope.

None of this is to demean hope per se. Hope is an important part of our humanity and is a well known motivator for all sorts of heroic and otherwise memorable deeds in the face of overwhelming odds.[14] But, despite their intimate relationship, faith and hope are not the same. In fact, that relationship can take on all sorts of values.[15] Faith can be accurate or not and the accompanying hope may be intense or weak. A range of combinations is conceivable and the interaction can be productive or destructive.[16] In general, however, the more accurate our faith the less prone we are to what can easily become an unhealthy reliance on hope. Indeed, rational faith can generate hopes with improved chances of being fulfilled.

Does Faith Just Happen?

Faith Happens is the name of a movie[17] and at least two websites[18] but is the sentiment correct? Earthquakes and hurricanes happen, cancer happens, accidents happen, and, lest we be too morbid, plenty of good things happen. For example, I recently spotted a billboard (for a casino) proclaiming, "Winning happens" (losing does too but losers lack the funds with which to erect billboards). So, does faith happen? Do marriages? Does getting a degree? Certainly all these things happen but I hope that, even in my short list, you detect the two fundamentally different ways in which they do. Some things, as we know from bitter or sweet experience, appear to be well beyond our control. Many others, however, seem at least partially subject to our thoughts and actions. Where does faith fit in this picture? In what sense does faith happen?

The strict determinist, obviously, believes that nothing is really within our control. From such a perspective, faith happens to us (or not) pure and simple[19] and even that belief (or lack thereof) is not our doing. I must con-

---

14. We might, however, question Petersen's claim that "hope shapes the future as much as anything else" (*Minding God*, 188). Peterson suggests that the basis for hope is "what we believe should happen" (ibid.). Perhaps this is frequently the same as what we want to happen (i.e., how I have defined hope), but although most of us probably believe that speeding motorists should be fined, our hope is that we, personally, are not.

15. When choice is involved, a sometimes useful way to consider the relationship is: motivation = faith + hope.

16. When the probability between two options is nearly even, the choice might be based on hope.

17. Garside, *Faith Happens*.

18. At the time of writing.

19. This seems to be the Calvinist perspective in (one line of) Christian theology but it is characteristic of any fatalistic viewpoint, religious or not.

fess that I don't have much use for strict determinism, as it provides little in the way of logical (or hopeful) suggestions as to why or how we might consider trying to progress as humans[20] but there are at least two senses in which it is correct to note that faith is outside our control.

In the first place, we have no choice about whether we will or will not have and use faith. Much of our discussion so far has emphasized emphatically that faith is not something we take or leave. Faith is not an optional cognitive accessory but is integral to who and what we are. In that sense, faith is beyond our control. Furthermore, as we noted in the chapter on faith and brains, both the operation of neurons and the status of the environment in which those operations occur are largely outside our control. Consequently, in many (if not most) circumstances, we sub-consciously form the probabilities that compose our faith. From this perspective, our conscious role in the process appears to be minimal. We may, for instance, occasionally be asked to characterize our belief about something and be surprised to discover that we can readily state and perhaps even defend a point of view even though we have never consciously considered the issue.

If that were the extent of the story, we might as well be zombies but there is more to say than this. In a very real sense we can play an active role in the pursuit of reasonable faith—that is, faith that can stand up to serious scrutiny or challenges to its validity. Every time we consciously determine to engage in a task that holds the potential to modify the confidence with which we hold a belief, we exercise some measure of control over how faith "happens" in our lives. From this perspective, aspects of our environments outside our direct control may yet be within our ability to influence indirectly. For example, we have little say over the facts, figures, and viewpoints presented in any specific lecture, speech, sermon, play, movie, or book but the choice to attend a particular college or church, to take this or that course from one instructor or another, to regularly watch a select movie genre, to read specific authors, or to avoid certain venues at all costs indirectly affect what will ultimately play a potentially significant role in shaping our beliefs. In short, there are important ways in which the quality of our faith is up to us. We can't help but have faith but we can help the faith we have.

---

20. If we think deity or destiny is manipulating our choices then what we believe, think, or do is already determined. A strict deterministic perspective, then, may not eliminate faith but it does make it superfluous.

The Quest for a Reasonable, Rational Faith

Presumably one could undertake the pursuit of an unreasonable, irrational faith but if we are content just to let faith happen, that will occur automatically with no special assistance from us. In any case, I will proceed in the hope (which, because I don't know you, is the best I can do) that you have more lofty ambitions. But be forewarned: I do not use the word "quest" lightly. Building a defendable faith can require significant commitment, beginning with the choice of where to concentrate one's efforts. Due to various limitations discussed in the previous chapter, we can't expect to attain the same high level of rational justification for all of our beliefs, and choosing focal points is itself an act of faith about what is of real importance. This is, of course, a dynamic process in which changes to a belief can cause us to reevaluate the perceived importance of the belief itself. Such feedback may lead us to divert our energies elsewhere.

The pursuit of rational faith is affected by a number of factors including our heritage, the quantity and quality of evidence involved, the approach we take toward such a task, and various obstacles that can impede our progress. Our goal in the following sections is to investigate each of these factors in order to gain insights that can contribute to a successful quest.

Recognizing the Context in Which Faith Forms

Despite its many positive benefits, what we can call "cultural inertia" is often to blame, both for the character of our thoughts and behaviors as well as our difficulty in escaping them when they are in error.[21] However, although acknowledging that one's heritage plays a significant role in shaping belief is important, just knowing it is not enough. Our hopes of achieving a rational faith with respect to any subject are contingent on keeping this constantly before us. The vehemence with which we often cling to cherished beliefs is testimony to the difficulty and reluctance we have in doing so.

I want to make it perfectly clear from the outset I am not suggesting that those cherished beliefs are wrong—only that they might be. Unless we recognize that, the quest for rational faith is stillborn. Furthermore, this is not a one-time concession. Even closely held beliefs that are the result of a long, careful process involving considerable research and reflection are

---

21. Consider: "Childhood acculturation strongly influences adult behavior; it shapes both our expectations and our sensory experience of the world in which we live." McNamara, *Evolution, Culture, and Consciousness*, 131.

acquired within some particular scientific, religious, or cultural paradigm. Paradigms themselves, however, are subject to change.

Several hundred years ago our ancestors learned that commonly accepted beliefs about the structure of the universe were wrong. Less than two hundred years ago sufficient evidence had accumulated to suggest that creation was an extended, dynamic process. It has been only a little over a century since people were faced with the knowledge that long-held beliefs about the absolute nature of motion, length, mass, and time didn't stand up to scientific scrutiny. Even more recently, behaviorism has yielded to cognitive science.[22] Changing views on slavery, gender roles, and ethnic equality are testaments both to changing beliefs and the difficulty associated with making those changes in socio-cultural contexts. The multiplicity of religions suggest that competing paradigms may exist for significant periods of time regarding matters of supreme importance, but even within any specific religion one can find diverse and changing views.[23]

Kuhn notes that new paradigms often take hold fully only after the proponents of the older paradigms have died off.[24] I hope you find this a sobering thought. His observation is undoubtedly true in areas other than science and applies to issues not large enough to count as paradigms. Most pointedly, it suggests that some folks are so bad at recognizing and correcting the grip of a flawed belief that only death can break its hold.

We often recognize the defect in others, asserting that they are stuck in a rut, narrow-minded, or old-school. At the same time we excuse our own blindness, calling it common sense or professional integrity, announcing that we will not yield to pressure, or merely protesting that you can't teach an old dog new tricks. I am inclined to suspect the beliefs of anyone who sees himself as an old dog but organ donation or some heroic act seem more profitable ways for one's death to contribute to the progress of mankind than the mere removal of one more instance of faulty and outdated beliefs.

For these insights to be useful, however, generalities will not suffice. To really appreciate the extent to which our beliefs are the product of the environments in which we find ourselves it is necessary to engage in specific

---

22. The classic look at paradigm shifts in science, including some of the ones mentioned here, is provided by Kuhn (*Structure*). Building on Kuhn's insights, Barker (*Discovering the Future*) shows how the paradigm concept can be extended to other domains and how easy it is to become so wedded to a paradigm that one misses significant possibilities.

23. The Protestant Reformation is a fine example. Barbour (*Religion and Science*) devotes an entire chapter to a discussion of similarities between models and paradigms in science and religion.

24. Kuhn, *Structure*.

personal analysis. This can be disconcerting but is a necessary first step along the path of rational faith. To get started, take a look at Figure 1 and complete the following sentence for as many of the areas shown in the cloud callout as you have courage to try: "If I had been born in _____ (place) in _____ (time), I would probably believe _____."

Figure 1. Contemplating the role of environmental context in shaping beliefs

For instance, you might say, "If I had been born in Ireland in 1824 I would probably not believe that humans could fly"[25] or "If I had been born in India in AD 300 I would probably believe in reincarnation."[26] Play around with different countries; contemplate ancient and modern eras. The exercise will be most beneficial if you are as specific as possible but don't feel too constrained by the template. The place and time of our existence are clearly crucial to the content of our beliefs but so are our gender, race, socio-economic status, and so forth. The offspring of aristocrats might, for example, believe that a monarchy is the most desirable form of government

25. Actually, this is tantamount to what William Thompson (aka Lord Kelvin), born in 1824, did seem to believe. Consider the following excerpt from his December, 1896 letter to Baden Powell: "I have not the smallest molecule of faith in aerial navigation other than ballooning or of expectation of good results from any of the trials we hear of. So you will understand that I would not care to be a member of the aeronautical Society" (Lienhard, "Baden-Powell"). Kelvin's view probably made it difficult for him to entertain any beliefs about frequent flyer miles.

26. Which, presumably, opens the door to a variety of flight possibilities not entertained by Kelvin.

while peasant children could conceivably be swayed otherwise, regardless of the country or era in which they live.

It is impossible to make a serious attempt to follow through with this exercise without squarely facing the prospect that beliefs underlying practices that the majority of humans currently consider abhorrent—slavery, cannibalism, kamikaze attacks, human sacrifice to appease some deity, plundering and pillaging—have been considered by others acceptable and in certain cases even sacred duties. Clearly there are some people who still see them so. Only disbelief in the power of environment can prevent us from admitting that, in another place and time, we could be those very people and that any immunity we think we have from such beliefs has its own environmental origin.

Believing that slavery, cannibalism, and the like are acceptable may seem far removed from our current beliefs but if you and I are, in principle, capable of believing any of these, what might we not believe?[27] If environment can exert so dominant an influence in matters of such consequence, all of our beliefs must surely be susceptible to its influence.[28]

The mental image I hope you are creating is of multiple hypothetical instances of you, a collection of virtual clones, each with a unique set of beliefs about what it means to be human, the source of meaning, the nature of God, how to find truth, the extent of human limitations, and also those myriad day to day circumstances and behaviors that receive the majority of our attention. One version of you votes Republican, another Democrat. Some versions have no chance to vote but believe that is the way things should be. Three instantiations of you are Hindu, five Islamic, and six are Christian. Several have no specific religious beliefs. Each individual embodies your beliefs as they would exist in a different time and place. There is one you that hates another you. Still another believes it is wrong to hate.

If this is too abstract, imagine that you and your spouse produce a large number of children who, immediately after birth, are shipped off to all

---

27. In his intriguing look at customs, many of which would appear to all but the practitioners as exceedingly peculiar, Montaigne asks: "What power does she (i.e., habit) not have in our judgments and in our beliefs?" (*Essays*, 79). Psychologist Richard Bentall, for example, notes that "people see ghosts because they believe in them" ("Why There Will Never be a Convincing Theory of Schizophrenia," in Rose, *Brains to Consciousness*, 132). For a humorous look at belief formation in American society, see Miner, "Body Ritual Among the Nacirema."

28. Kaufman refers to "a world-picture that will be largely taken for granted in all future acting, thinking, planning, exploring, meditating, and ongoing living" (*Jesus and Creativity*, 82). Of course I am hoping (and Kaufman too, I think) that we can see the extent to which we take things for granted and find the wherewithal not to do so in such a perfunctory manner.

corners of the globe. For many years they are raised by surrogate parents, educated and otherwise steeped in the culture in which they have landed, but having no contact with you. When you eventually decide to hold a family reunion, what can you expect to find? Well, we all know what to expect because it would be quite close to what we actually observe with the large number of infants from other cultures who are adopted by U.S. parents and grow up sharing the basic belief systems prevalent in their new environment.

Among the myriad possibilities for beliefs, each of us has materialized in a specific environment that promotes the acceptance of one particular sub-set. It is inconceivable that all of those beliefs we call our own are correct. Recognition of this fact can entail two quite different responses. One is to worship at the altar of cultural relativism, exalting the diversity of beliefs and thinking that, in some fundamental way, they are more or less equivalent or at least that there is little hope for distinguishing their relative merits. The alternative is based on the assumption that all beliefs are not equally meritorious. This approach embraces the search for truth and acknowledges that environment is not the whole story. It concedes that we cannot escape our environment but that we can learn how to interpret it. Both approaches are themselves beliefs, certainly, but whereas the first suggests resignation, the latter offers promise that we can discover a justifiable sub-set of beliefs (including beliefs about what really matters).

The Importance of Evidence

Imagine that you have been summoned to jury duty. Here are the prosecutor's sole remarks:

> Your honor, the defendant is undoubtedly guilty. He has the look and demeanor of a scoundrel. I feel it, your honor! In all my experience there has never been anyone whom I believed more clearly to be at fault. Why, last night, I even dreamed that he was guilty. Take my word for it, he's the culprit. What more need I say?

I daresay you would be astounded if your fellow jurors voted to convict on the basis of those remarks. How could there be no reasonable doubt in their minds? Where is the evidence? You may question whether the prosecutor passed the bar, perhaps even feeling confident that he stopped at one on his way to trial because, without evidence relating the accused to the crime, there is no way you (or your peers) would consider him guilty. A juror's assignment is to arrive at a conclusion about the truth of a matter and those bogus legal comments above strike us as bizarre precisely because they neglect the evidence we expect and consider essential for doing so. God may have explained to the lawyer that the defendant was guilty but unless we overheard that conversation we would expect some other kind of corroboration. Unfortunately, what seems so obvious in court often escapes us elsewhere.

Consider the tendency to discount the need for evidence. We learn in high school physics (or earlier) that, in a vacuum, falling objects accelerate toward the earth at the same rate. Yet if Aristotle had been our teacher, we would have been instructed otherwise.[29] We might well wonder how such a sage could get it so wrong but Galileo, who helped rectify the erroneous belief, made it clear: "I greatly doubt that Aristotle ever tested by experiment whether it be true."[30] The evidence was there for the taking but was left untouched, not only by Aristotle but, apparently, by everyone else for the next nineteen hundred years. Did they, like our cartoon lawyer, merely "feel" the truth of their beliefs?

I hope it strikes you as a matter of some concern that if distinguished Greek philosophers can appear to have disregarded the need for evidence it does not bode especially well for the rest of us. The tendency of all those people living in the period between Aristotle and Galileo to do so was no doubt partly due to the context Aristotle bequeathed them, but blaming others for our own failures to look for or consider the importance of evidence does little to solve the problem. It is not that most of us make a conscious decision to overlook evidence, although that can happen. The fundamental problem is that we usually just fail to make any conscious efforts not to ignore it. Such disregard is a passive byproduct of the failure to fully appreciate its importance.

Francis Bacon was an early and persuasive spokesman for the role of evidence in the sciences, as well as the need to handle it well. He looked to nature, first and foremost, for the evidence he saw to be essential to its proper interpretation,[31] a sentiment echoed eloquently several hundred years later by the theologian Henry Drummond:

> The danger of philosophy putting in the ends is that she cannot convince everyone that they are the right ones. And what is the valid answer? Of course, that Nature has put in her own ends if we would take the trouble to look for them.... The philosopher

---

29. Aristotle, *Physics*, book IV, Part 8.

30. Galilei, *Two New Sciences*. As Ferris notes, "We can live by dogma or discovery" (*Science of Liberty*, 261).

31. Bacon, *Great Instauration*, Preface.

requires fact, phenomenon, natural law, at every turn to keep him right; and without at least some glimpse of these, he may travel far afield.[32]

I suspect that most of us deem many of our other convictions to be at least as important as what we think about nature, so why we would be satisfied with any less rigor in those non-scientific beliefs is something of a mystery.

Even when the evidence is clear, we may choose not to follow it—sometimes with serious consequences. Confirmation of this disrespect for evidence is all around us. Though few argue with the validity of the Surgeon General's warning about smoking, the perils of texting and driving, or the odds of successful gambling, lives are frequently ruined because people behave as though there was no supporting evidence.

But this is not only a matter of picking and choosing which evidence we will consider and which we will not. Sometimes it simply involves settling for something inferior. For example, in deference to his otherwise significant contributions to our thinking, one might be inclined to give Aristotle the benefit of the doubt and believe that he merely relied on inferior evidence for his conclusions about the effects of gravity. Yet, the simplicity of Galileo's demonstrations suggests that any such "evidence" couldn't have been too substantial. The best we can say for Aristotle in this case is that, despite the flaws in his thinking, at least he was doing so.

It is easy, however, to deceive ourselves about the depth of the thinking we do. This is one of the dangers of rationalization. As the handy servant of wishful thinking, rationalization can frequently be seen as a sad attempt to justify a belief at all costs. Although that is far from rational, there is a cure. It is called evidence.

However, even when we appear to be thinking, our beliefs can take a prearranged path. Many times we follow the evidence to a foregone conclusion—a destination to which we have determined ahead of time it will take us. This is part of the problem with theories that are not "falsifiable," a term popularized a number of years ago by philosopher of science Karl Popper in his attempts to characterize legitimate scientific theories.[33] A theory that can be falsified is one for which it is possible, in principle, to show that it is wrong. If there is no conceivable evidence that could falsify the theory, according to Popper it cannot be labeled scientific. Consequently, Popper considered Marx's economic theory and Freud's psychoanalytic theory non-scientific because any observations one might make can be worked into the framework

32. Drummond, *Ascent of Man*, 19.
33. Popper, *Philosophy of Science* in Mace, 155–91.

of their hypotheses. There is no way to prove them wrong, not because they are necessarily incorrect but because they are too general—they take no risks. Popper is careful to note this does not mean they are in fact wrong—he believed there was merit to Freud's ideas, for example—but that their inability to stand up to the falsification criteria saps them of any claim to scientific legitimacy. A genuine scientific theory, in contrast, makes predictions that could, due to some observation (e.g., perhaps via experiment), be undermined. Even Aristotle's hypothesis about falling objects was falsifiable, though it took hundreds of years for the necessary observations to be made.

When a theory is not falsifiable, it entails the adoption of a foregone conclusion. As a result—as Popper noted[34]—anything subsequently considered from the perspective of that theory will only appear to confirm it. Even in theories deemed falsifiable it is possible to insert components that are not. For example, it is easy to attribute the bright red plumage of a male cardinal to its adaptive evolutionary advantage for attracting potential mates. That perspective, it then seems, makes the transitory coloration of the chameleon appear problematic. How is it to be found by its would-be suitors? On the other hand, the ability of the chameleon to match its color to that of its environment can be considered a protective adaptive evolutionary advantage that enables it to avoid predators. But doesn't the bright coloration of the cardinal attract its enemies? The problem here is not with the theory of evolution in general[35] but in making it too general. If the theory was nothing more than a vacuous truism that survivors survive, it would have little to offer in the way of useful explanation and practically anything could be made to fit its preconceptions.

But I am not primarily concerned here with scientific theory alone. I have used this example because what can occur in association with a presumably rigorous discipline such as science can easily take place in other areas of our thinking where we are subject to interpret things in the darkness of our foregone conclusions. Even the very evidence that could and sometimes should encourage us to revise our perspectives is all too easily worked into the framework of existing beliefs.

Mark's account of Jesus healing a deformed man[36] suggests that medical intervention of that nature might meet with some sense of awe, espe-

---

34. Ibid.

35. The discovery of a (presumably fossilized) pre-Cambrian rabbit is supposed to be the falsification criterion for evolution suggested by evolutionary biologist J. B. S. Haldane. (Charlesworth and Charlesworth, *Evolution*) Actually, it would only falsify certain aspects of the theory but the point is that key components are falsifiable.

36. Mark 3:1–6. I alluded to this incident earlier when discussing misconceptions about faith.

cially when the physician is not board-certified. That the observers used the event as incentive to plot Jesus' death indicates that when one is "looking for a reason"[37] to support an existing mindset, it is difficult for evidence, no matter how dramatic, to have much impact.[38] Stubbornness may have redeeming qualities in some contexts but not when it extends to the exclusion of evidence.

No wonder Jesus was distressed. That is a typical reaction when we see anyone attempt to make evidence fit their foregone conclusions to the exclusion of a more logical interpretation. No wonder Galileo had little patience with those who hoped the new evidence he provided could somehow still fit their old ideas about the appropriate structure of the cosmos. No wonder biologists are mystified by an obstinate refusal on the part of some to even consider how biological and geological evidence support evolution better than it does other ideas. Yet none of us are immune. Idealization of a particular type of government leads to a tendency to justify whatever it does. Lovers can deem otherwise irritating quirks endearing. How we interpret the biblical story above will reflect preconceptions about its credibility. Our default operating mode, apparently, is to make evidence fit the belief rather than make belief fit the evidence. But this is not the road to rationality. Might we hope for something better?

Syndicated columnist Cal Thomas, writing in the wake of celebrated atheist Christopher Hitchens's untimely death, remarked that, "Evidence alone has never moved anyone from unbelief to faith."[39] Clearly, as we've been noting, evidence can be ignored or minimized, but has it "never" been responsible for a change in one's belief? To be fair, Thomas's statement refers to "evidence alone" so there is a sense in which he must be right because there is never any such thing as "evidence alone." Anything we call evidence must always be apprehended in some context and will be processed in light of our wishes and desires. But, if evidence cannot do it, what can? It is not likely that the absence of evidence will substantially raise or lower the probabilities associated with any given belief (although it will not prevent one from holding onto one that is ill-founded). If evidence alone couldn't change

37. Ibid., 2.

38. According to Mlodinow, "When we are in the grasp of an illusion—or for that matter, whenever we have a new idea—instead of searching for ways to prove our ideas wrong, we usually attempt to prove them correct. Psychologists call this the confirmation bias..." (*Drunkard's Walk*, 189). Mlodinow goes on to quote Bacon on this (Bacon, *Great Instauration*, "Novum Organum," aphorism XLVI).

39. Thomas, "Hitchens Smart." Thomas thinks that it is God's gift of faith that makes it possible for some people to believe in him. This is not the place to review the theological issues of freedom but Thomas might also want to consider that Jesus suggested to his disciples and others that they look to the evidence (cf. John 14:11; 10:37–38).

a belief, why did Jesus tell a doubting Thomas (not Cal!) to put his hand into the hole in his side?[40] If evidence is not central, the bachelor trying to convince his girl to marry him can safely quit proclaiming and showing his love, secure in the knowledge that she will just know it.

In summary, faith must ultimately be verified if it is to have any real merit. Evidence is the source for that confirmation and forms the basis for sound probabilistic reasoning. It is the means by which we avoid the problems associated with blind faith. An evidence workout is the primary means of firming up what Montaigne calls "softer" minds,[41] which are prone to think without confirmation and to focus on what could have been versus what was, or on what might be versus what is.[42]

What Counts as Evidence?

Even when we accept the importance of evidence, are driven to pursue it, and work hard not to follow it to a foregone conclusion, we are still faced with the task of deciding whether some observation counts as evidence. Should we believe a manufacturer who makes certain claims intended to convince us to purchase a particular product? Is a successful senatorial career evidence that a person will make a good president? Can we count on the veracity of a witness who is closely related to a defendant? What can make a defendant's own testimony dubious? Why is it that courts frown on hearsay or information obtained by illegal means? I suspect it is clear by now that ulterior motives, poor application of logic, wishful thinking, and taking things out of proper context can all affect the credibility of something that might otherwise count as evidence.

However, even if something is deemed evidential, we are still frequently faced with the task of deciding toward what specific belief that evidence should apply. Is the fact that you got sick one evening after eating seafood evidence for an allergy, an incidental stomach virus, or poorly prepared food? Does a knocking in your car's engine signal imminent failure or does it just mean there is some water in the gasoline? Is a rash of airline crashes evidence that there are problems with pilot training, the aircraft control system, maintenance of the planes themselves, or terrorist activities? Are

---

40. John 20:24–29.

41. Whose "belief has been so strongly seized that they think they see what they do not see" (Montaigne, *Essays*, 70).

42. The problems identified in this section plague many religious beliefs but are also manifest in other domains of thinking, including science (cf. Lindley, *End of Physics*, 20).

weeds in my yard evidence of a disregard for horticultural esthetics or an indication that I spent the time during which I could have been mowing thinking about examples that indicate how something might be evidence for one belief but not another?

The significance of information is always subject to alteration by additional information. For example, poor marks in school are evidence of something, but what? Is the student incapable of handling specific subject matter or just poorly prepared in earlier classes? Did illness, loss of a family member, or a broken relationship create a set of extenuating circumstances that made it impossible to concentrate on the coursework? Knowing that the student had high scores on standardized tests (e.g., the SAT) might suggest that the low grades were a passing aberration. On the other hand, we might learn that the student has a playboy mentality or is addicted to computer games. Then we would be less inclined to think his poor performance would not be repeated.

The bottom line is that few things, if any, count for evidence in isolation. We have already noted the importance of quality but the quantity of evidence is no less significant. The creation of sound beliefs depends upon collecting as much evidence as possible. Yet, even volumes of information collected from a single source can be suspect. Consequently, justifiable beliefs must almost always also rely on evidence from multiple sources. Obtaining the truth, the whole truth, and nothing but the truth entails consideration of the whole body of evidence (and nothing but the evidence). The more observations we make and information we obtain and the more sources we use, the more likely it is that we will detect contradictions between things that we might otherwise count as evidence. This process also improves our chances for identifying credible sources of evidence.[43]

As we saw earlier, faith always forms in some context, so it is not surprising that the things we are willing to count as evidence also depend upon context. In the world of science, the veracity of scientific statements is accepted to the extent that those statements are capable of being directly and repeatedly verified. This is the context the scientist is accustomed to expect. Historic events, however, are by definition non-repeatable but that doesn't prevent us from sometimes accepting historic statements as evidence of the events they are meant to describe. The historian's criteria for acceptable evidence will have many of the same attributes as the scientist's, but repeatability and (except in limited cases) direct observation are not among

---

43. Nobody is claiming this will be easy. The next time a drug is advertised on television, pay close attention to the disclaimers. All those potential side effects constitute part of the overall body of evidence that must be considered in deciding whether the drug will really be beneficial.

them. And, despite the fact that history doesn't repeat itself, historians usually expect that any alleged evidence will have a similar character to that accompanying comparable events. This is the context the historian is accustomed to expect.

Both scientists and historians are interested in the pursuit of truth but when things don't fit the relevant context, there is less willingness to accept them as evidence. One of the reasons that there are still misgivings in some quarters about evolution of species and certain cosmological theories is because there is no known way to provide direct verification—conclusions must be formed via inference. If the resulting evidence is more palatable to the historian than the scientist, it is easy to see why.

Despite all this, scientists are willing to count the testimony of fossilized animals or the witness of background radiation and red shifted spectra of receding stars as confirmation of their beliefs (and there are good reasons for doing so). With comparable need the historian acknowledges that, just because an event is extraordinary, the evidence for it need not necessarily be less genuine (although that is sometimes the case). Both concessions betray a willingness to extend the preferred criteria for what counts as evidence when circumstances are deemed to warrant it, but that decision itself is impossible unless one first determines that there is evidence in its favor.

These considerations are important when it comes to deciding what counts as evidence in exceptional circumstances. Most religions, for example, have components that, because they are historic, are subject to the same types of evaluation that accompany other purported events of the past. What makes many religious assertions problematic is that the types of things claimed are frequently outside the normal realm of experience. This would be what Bierce meant by "things without parallel,"[44] something I'll address in a later chapter. However, although every event is unique and non-repeatable, we often focus on event attributes categorically when evaluating the probability that a reported event is factual. Believing that Washington crossed the Delaware is well within the scope of our usual experiences but we may find it more difficult to accept religious visions (or, say, relativistic accounts of length contraction) because they are not. Reports of events that fall somewhere in between (e.g., walking on the moon) may be met with various degrees of skepticism.[45]

---

44. Bierce, *Devil's Dictionary*, 40.

45. One of the reasons people like science is because verification is sometimes relatively easy. However, although the scientific approach provides a worthy standard for the evaluation and acquisition of evidence, it maintains its aura of evidential invincibility in part by a hesitancy to even consider certain matters. Yet many of those concerns are quite important and we may be unwilling to ignore them (cf. Robinson, *Absence of*

None of this is intended to suggest that what counts as evidence for religious claims should be immune from the general rigor expected in other domains. However, it seems only fair that the same concessions we grant to science and history should also be extended to religion. How far we should go requires careful consideration. For example, the intensely personal feelings that accompany religious experiences are hard to verify for any but the practitioner but this is also true of most of the phenomenal aspects of consciousness. Despite this, we are quite willing to count reports of sensations and emotions as evidence that some phenomena has occurred.[46] Of course, it is one thing for Sam to believe that Sue had a religious experience and quite another to count it as evidence for the religion itself. But consider that if Sue tells Sam she feels she is in love with him, he will be inclined to interpret that as evidence that she really is (and not merely as evidence that she had the feeling).

One thing is clear: something cannot count as evidence for a particular belief if that belief is thought to be false. This presents us with a Catch-22 situation where we need to know whether to count something as evidence that supports a particular belief but until we know that the belief is true we don't know that we should do so. Given the recursive nature of belief formation, however, this should not surprise us. For example, someone tells us that there are no fish in a pond. Should we count that as evidence that there are not? Surely we will consider the source, ask other people, and seek other evidence but, if we throw in a hook and line and pull out a fish, we will know for sure that the pessimist's statement was really no evidence at all. If we fail to catch a fish, however, we must continue to wonder about the veracity of the claim—perhaps the fish weren't hungry.

The Role of Assumptions

I once spent a significant part of a long day wandering around on a glacier in the mistaken belief that I was on my way to the highest peak in Wyoming. That was many years ago but I have wondered on a number of occasions since then how, topographic map in hand, I managed to miss the ascent to the pass that would have taken me to my destination. The route I chose certainly looked like I thought it should. Whatever the reason, it is probably safe to say that I was enticed by one or more faulty assumptions (not the least of which must have been related to my map-reading skills at the time).

Mind; Ross, *More Than a Theory*).

46. Dennett (*Sweet Dreams*) suggests making something like this an integral part of a scientific approach ("heterophenomenology") to studying consciousness.

But each of us routinely wanders through a bewildering space of beliefs.[47] Yet as we saw in the last chapter, faultless logic can only take one so far—beliefs rise and fall with the quality of the assumptions on which they are founded. Galileo saw this clearly as he promoted his view of the cosmos in the following fanciful dialogue:

> Simplicio: Aristotle gives a hundred proofs that the universe is finite, bounded, and spherical.
>
> Salviadi: Which are later all reduced to one, and that one to none at all. For if I deny him his assumption that the universe is movable all his proofs fall to the ground . . .[48]

Obviously, Galileo had no choice but to employ his own set of assumptions. As Maritain has noted, "every science, except the highest, bases its demonstrations on postulates or data it is incapable of explaining or defending. For instance, mathematics does not inquire what is the nature of quantity, number, or extension, nor physics what is the nature of matter."[49] The highest science for the philosopher Maritain was not, sadly, physics or even computer science, but (surprise!) philosophy. But isn't Maritain's statement itself an assumption? Even philosophy needs postulates, premises, and hypotheses. Yet philosophers need not feel badly about this—faith of every form and faith in anything, from theories of the cosmos to the choice of the most palatable entrée at a restaurant, requires assumptions.

Because many of our beliefs are based on the authority of others, we should acknowledge that our trust in that authority also involves assumptions. Many of those assumptions are quite natural, perhaps even warranted, but they are not always valid. This can mean that, sometimes, our beliefs are incorrect because the sources we trust to inform those beliefs are themselves off track. Consider this example from a National Geographic tome entitled (unfortunately, in this case) *The Knowledge Book*. See if you can spot the problem:

> 51 Pegasi is the first sunlike star that was discovered to have an orbiting planet . . . Its distance from the star is equal to 20 times the distance between the Earth and the sun. Therefore the planet may be as hot as 1832°F (1000°C). It takes only 4.2 days

---

47. Taking a cue from the old Christmas hymn, we might do a bit more wondering as we wander.
48. Galilei, *Two Chief World Systems*.
49. Maritain, *Introduction to Philosophy*, 72.

to complete an orbit around the star ... 51 Pegasi is comparable to the sun in mass and size.[50]

If you are asking yourself how a planet that far away from its star could it be so hot or wondering if its rate of revolution should be much less than that given, you are on the right track. Apparently the correct distance is more like 1/20th of that from the earth to the sun,[51] a figure of which you might also be suspicious (but at least it makes sense in context of the other values). Certainly it is easy to attribute this to a simple typographic error but that would be to miss the point. Although the source of a false belief may be carelessness, ignorance, or deliberate fraud on the part of an authority we have assumed we could trust, it behooves us to be mindful about our assumptions regarding that authority.

There is something more important in Maritain's remarks, however, than a basic observation about the need for assumptions. Although it may have once been true, particle physicists of today would likely dispute Maritain's assertion that physics does not inquire into the nature of matter.[52] I don't mention this to criticize Maritain—he made his observation many years ago—but to emphasize the hesitancy we should have, not just about accepting an assumption but even about accepting that something is (or should be considered) an assumption. Such an attitude was part of Bacon's vision for the proper conduct of science: "I also sink the foundations of the sciences deeper and firmer; and I begin the inquiry nearer the source than men have done heretofore, submitting to examination those things which the common logic takes on trust."[53] In the case of assumptions, what is good for science is good for every area of our thinking.

How Much Evidence is Enough?

Lest we look like blind men attempting to describe an elephant,[54] it is helpful to keep in mind how the various factors already considered affect any reasonable answer to this question. We could let it go at that but there are other useful ways to contemplate whether our beliefs are sound. Even when standards for evidence are high, the truth is that any amount of evidence that

50. Grogan, *Knowledge Book*, 17.
51. Cf. http://www.exoplaneten.de/51peg/english.html.
52. The large hadron collider in Switzerland provides compelling evidence for this counterclaim (cf. Radowitz, "Back to the Beginning").
53. Bacon, *Great Instauration*, "Arguments of the Several Parts".
54. Saxe, *Poems*.

pushes the probability associated with a belief past 0.5 is probably sufficient to make us behave as though we really do believe it. Of course, that doesn't mean we should. The key is to distinguish between how much evidence is enough to make us believe that something is true whether it is or not, and how much is necessary to make us believe that something is true which, in fact, really is true.

Consider, for example, that you are trying to determine whether to believe that some particular event occurred. Unbeknownst to you, there are seven pieces of evidence indicating that the event did occur and three suggesting that it did not. We'll call the evidence in the first category confirming evidence and that in the second, disconfirming evidence. For the sake of further simplifying the example, imagine that all evidence carries the same weight. Now, how much evidence is enough to make you believe that the event did or did not occur?

In the absence of other factors and if you had no confirming evidence, even a single piece of disconfirming evidence could be sufficient to convince you that the event never happened. In fact, as long as you believe that the ratio of disconfirming evidence to confirming evidence is greater than 0.5, you will be inclined to also believe the event did not occur, even though (in our example) there is plenty of evidence available to suggest that it did. The overwhelming evidence for the occurrence of the event is not enough if that evidence is not in your possession or if it is ignored.

Let's define the "belief probability" for this example as the probability that the evidence in your possession will be sufficient for you to believe what the confirming evidence is supposed to support. This is a function of the confirming and disconfirming evidence.[55] If we plot this function for our example (as in Figure 2), we can visualize how the belief probability changes with various combinations of evidence types. Points on the plotted surface having belief probability values above 0.5 (indicated by a plane halfway up the vertical belief probability axis) represent cases where there is more confirming than disconfirming evidence. In this contrived example, where the pool of confirming evidence is greater than the pool of disconfirming evidence, there are obviously more ways in which the belief probability can be biased in favor of the belief in question than there are ways in which it can be biased against the belief but this need not be the case in general.

---

55. Belief probability = confirming evidence / (confirming evidence + disconfirming evidence).

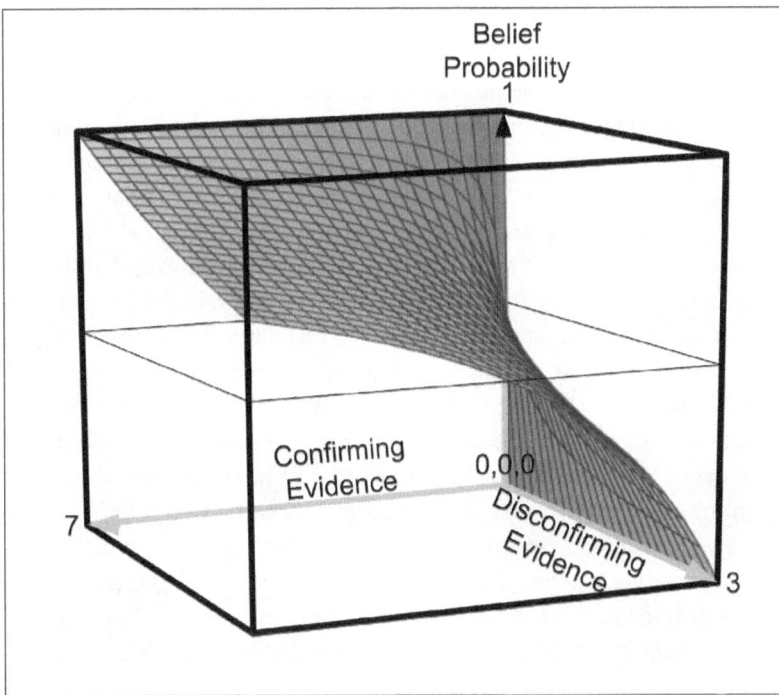

Figure 2. Belief probability as a function of accepting various combinations of confirming and disconfirming evidence. In the plot shown here there are seven possible pieces of confirming evidence and three possible pieces of disconfirming evidence.

Obviously this is a grossly oversimplified example[56] but the same issue arises in real life—it is just more difficult to see. Yet as bad as it is to fail to possess available evidence, things can be worse. With a predetermined desire for something to be true, even a preponderance of available disconfirming evidence might go unheeded. We have all run across situations in which it seemed apparent that no amount of evidence would be adequate to cause a change of mind. Unless that mind is our own, such situations are usually puzzling and we wonder how anyone could be so obtuse. If the obstinacy belongs to us, however, we can be quick to excuse ourselves—we don't acknowledge or believe that any contrary evidence really exists, we cannot imagine any other possibility, and so forth. Perhaps we are correct. It is less convenient to concede that we might be mistaken on all counts; that we are so closed even to the possibility of being wrong that our only "rationale" is feeling or fear or wishful thinking. In such a case, when no

---

56. Most obviously, weighting all the pieces of evidence equally.

amount of evidence is enough, it is time to reevaluate the belief. From a practical standpoint, we need only be concerned to do so in matters of real importance. Yet, as we've already seen, making that determination (i.e., what is really important) might be the very belief that requires reevaluation.

Due to our limitations, of course, there will seldom be enough evidence to give us absolute (as opposed to partial) certainty about most beliefs. Knowing this is vital to a proper understanding of faith and knowing what to expect from evidence. Theologian Keith Ward provides a nice account of how this works with regard to belief in God by remarking that even though we cannot obtain "total theoretical certainty" in the existence of a particular type of God, we "can have good reasons for thinking there is such a God."[57] Someone else might wish to say that we can have good reasons for thinking there is not but the belief itself is not of concern here. The function of evidence is to help us get the probabilities right and a belief is justified when the evidence does so.[58] That is how much evidence is enough.

I've noted before that we can never even be certain of something so basic as whether a particular individual really loves us. However, we can still be justified in believing they do. We submit as evidence their words of affection, remarks of praise to others, caresses, gifts, and sacrifices on our behalf. We point out the high frequency with which those signs appear and the quality of each one. We also notice the lack of evidence which would suggest that the gestures are not merely a pretense based on some desire to maximize the inheritance or keep receiving financial support. Although we acknowledge the acting skills of others and are aware of the potential for our own gullibility, the gradual accumulation of such evidence continually raises the probability that our belief is sound. But getting as much evidence as we need is not the same as getting as much as we want. Neither is it essential that we get all we want. Rational faith is a function of how we handle the evidence we do manage to get.

In scenarios like the one above, however, we only have enough evidence for the moment. If at any time we become aware that our "loved one" has taken out a large life insurance policy on us or sold our toys while we are away, we might be inclined to change the probabilities of our belief. Deciding when we have enough evidence to do so is a dynamic proposition and requires a dynamic mindset. It is the mindset of an explorer.

---

57. Ward, *Divine Action*, 14.

58. Ibid. Ward's claims regarding what we "cannot be justified in believing" seem to be misstated in this regard.

––––– CHAPTER 8 –––––

Explorers and Mechanics: Creative Approaches to Faith

We only ascribe universal education to one who in his own individual person is thus critical in all or nearly all branches of knowledge, and not to one who has a like ability merely in some special subject.[1]

—ARISTOTLE

ACCORDING TO ROBERT BENCHLEY, "There may be said to be two classes of people in the world; those who constantly divide the people of the world into two classes, and those who do not."[2] Well, I normally teach three classes each semester and know plenty of other professors who do so as well. That adds up to several hundred classes at my university alone, so you can see why I am somewhat skeptical of Benchley's conjecture.[3] In any case, Benchley's statement alone has generated two additional types of people in the world: Those who have heard it and those who have not. But, among those who

---

1. Aristotle, *On the Parts of Animals*, book I, part 1. In an era of exponentially increasing knowledge the person described by Aristotle seems inconceivable. However, it is better to see this as a target which enables possibilities for forming valid beliefs. It is also a warning that, although all beliefs should be held cautiously, it is particularly important to do so in areas where we lack competence (one of which might be the ability to judge our own competency).

2. Benchley, *Of All Things*, 187.

3. You may be thinking that there are those who know when to leave attempts at humor to the professionals and those who don't.

have heard it, there are those who think Benchley was correct and those who do not. And so it goes.

Benchley was poking fun at our tendency to pigeonhole one another but this is not something some of us do and some don't. Although we might each engage in the process to varying degrees, categorization is an integral and necessary part of our cognitive processes. Yet, because it is so natural, it is easy to get carried away and our classifications can end up as expressions of an irrational bias that compress people into predefined molds from which we seldom allow them to escape. At this point classification has lost its utility and is harmful to both the classifier and those classified.

I am hoping that recognizing this danger will keep us on our toes as we tiptoe through this chapter because, despite my general hesitancy to overtly categorize folks, I intend to proceed here by doing just that, dividing the world into two kinds of people: explorers and mechanics. The justification for employing this classification scheme hinges on several factors. First and foremost, it provides a useful approach to thinking about how we bring reason to our faith. Second, although experience suggests that some people are more or less permanently in one of these categories or the other, that need not be the case. In fact, many of us will fall first into one, then the other, alternating between them at various stages in our lives. The amount of time we spend in any one category can also vary. Finally, although our natural tendency to classify things will cause us to judge to which of these categories we believe others belong, the simple fact is that we put ourselves into one or the other—I have merely provided the labels. Ultimately, the thoughts and behaviors of each individual determine if he or she is an explorer or mechanic. Consequently, these are categories from which we can enter or leave at will and our foremost concern should be about our own classification and not that of someone else.

When we think of explorers and mechanics, we are apt to get fairly vivid mental images of each. That is one reason I use these particular categories. One of my favorite explorers, for example, is John Wesley Powell, the one-armed Civil War veteran who led expeditions into little known regions of the American West in the nineteenth century, including a traverse of the Grand Canyon by boat.[4] Undaunted by his physical limitation, lack of whitewater skills, or the questionable credentials of his crew, Powell proceeded to do what innumerable explorers before and after him have done—to boldly go where no one had gone before.[5] From Columbus to

---

4. Cf. Powell, *Exploration of the Colorado River*.

5. In Powell's case the Anasazi and possibly others had preceded him but if any did so by boat there is no record of it.

Captain Kirk, explorers have been, are, and always will be dissatisfied with their current level of knowledge (and possibly that of others as well). Driven by curiosity, they are not content to sit at home (or anywhere else, for that matter). Explorers embrace the unknown, suppressing their natural inclinations for comfort and safety. For them, risk-taking is not a necessary evil but part of the joy of discovery. Explorers are resourceful and creative, and their intense determination and initiative make them not merely active but proactive. The explorer proceeds not knowing what he will find but confident that he will find something worth knowing. The antithesis of this mindset is derided thus by the philosopher Robert Spitzer: "I do not have to seek an answer; I don't even have to ask the question. I can behold incomplete intelligibility and instead of pursuing its invitation, eat a bon-bon and watch a rerun on television."[6]

My reference to Spitzer's satirical remark should make it clear that the kind of exploration I have in mind is not limited to that of the physical universe. It is the set of attributes that drove a Lewis or a Clark that concern us, whether they are applied toward discovery of new lands, new scientific theories, new creative expressions or inventions, or just new ways of understanding. It is the *mindset* of the explorer that is essential to the development of rational faith.

Now that you have a mental image of the explorer before you, contrast it with the mechanic. This should be easy, because we encounter them as often, if not more so, than we do explorers. They are not usually as flamboyant, but when something breaks or we feel the need for preventive maintenance, they are indispensable. We take our vehicles to them, call upon them to repair appliances in our homes, and rely on them to maintain the machinery that supports critical parts of our infrastructure such as power generation facilities. The role of the mechanic is essential but the operative terms are "maintain" and "mechanical." Mechanics fix things that already exist. They maintain the status quo. When my wife takes our twenty-year-old van to the shop, she doesn't return with a new one but with the same old van plus a few additional patches.

In addition, repair or maintenance procedures themselves are relatively automatic. This is not to say that it doesn't require significant skill or experience to be a successful mechanic but that the diagnostic process follows a predictable line of reasoning: if feature A is malfunctioning then check feature B otherwise check feature C. Both repair and maintenance are grounded in a regimen that has proven workable in the past.

---

6. Spitzer, *New Proofs*, 264.

Mechanics, however, come in a variety of flavors. Physicians, teachers, scientists, and corporate professionals alike operate in their own mechanical worlds, following largely reflexive procedures to keep their patients, classes, theories, or businesses running. But when methods of medical diagnosis or the usual treatments fail to work for strange illnesses, when teaching strategies no longer reach a certain student population or are ineffective for presenting new material, when theories cannot handle novel data, or when business climates change and old techniques stop working, it is not sufficient to be a mechanic. If you want to keep an old car running, you will call a mechanic. If repairing the old one is no longer a viable option, you will have to do a bit of exploring.

The mechanic's approach is also just the ticket for keeping old beliefs operational. But if those beliefs seem to have lost their rhyme or reason; when they are no longer able to stand firm in the face of serious scrutiny; when attempts at patching them rip their fabric further, it is time to become an explorer.[7] That is the fundamental difference between explorers and mechanics—mechanics preserve the current state of affairs; explorers are instruments of change.[8]

Both have their place, certainly, but identifying just what that is may not be easy. A crisis of faith doesn't necessarily mean that an old belief is unsound. Like a vintage automobile, many beliefs are well worth preserving and even parading, being in no danger of losing their serviceability or value. But there are also cases where safeguarding the old beliefs is far from safe; where a complete overhaul or total replacement is the only rational course of action. The mechanic in us, however, may never see this. Unless we are in the habit of at least periodically adopting the explorer's outlook, we will view our charge as maintaining, even protecting existing beliefs while simultaneously defending the mechanic's approach against all comers. Only the explorer knows when it is appropriate to be a mechanic.

The real explorer also recognizes the need to investigate the concept of exploration itself and understands that search alone is of little value. This simple fact is implicitly acknowledged but overtly denied by many who persist in exalting the journey and viewing the search process as its own end. But what good is a search without discovery? The whole point of the journey is to see new sights and learn new things—in the arena of belief to establish a firm rational foundation. Without such productivity, however,

---

7. This is the scenario Kuhn (*Structure*) describes as characteristic of a scientific paradigm ripe for change and parallels the insight of Jesus about the mistake of trying to put new wine into old skins (Luke 5:36–39).

8. Changes implemented by mechanics are primarily to preserve a previously existing condition.

the glamour and expectation of exploration disintegrate into a hopeless, perpetual wandering among alternative beliefs that look equally dull and unpromising. The would-be explorer who lacks a home base, who finds nothing of promise along the route, and who establishes no outposts with potential for long-term viability, is as tightly held by a unique grip of routine and convention as the most resolute of his mechanistic peers.

The tension between automaticity and exploration shouldn't surprise us since it is typical of the way our minds work in general. Often it is only by concerted, conscious effort that we can identify our unconscious biases, drives, desires, and other robotic responses and tell them where to get off, and even then it may only be with great difficulty. That which makes it feasible to think that our beliefs can move in a positive rational direction despite the trouble it causes is recognizing that being an explorer is a conscious decision—our conscious decision—and then resolving to be one. Yet, it is not clear that this is something at which we are very good. Consider, for example, C. S. Lewis's perspective on belief during the Middle Ages:

> I am inclined to think that most of those who read "historical" works about Troy, Alexander, Arthur, or Charlemagne, believed their matter to be in the main true. But I feel much more certain that they did not believe it to be false. I feel surest of all that the question of belief or disbelief was seldom uppermost in their minds.[9]

But perhaps it should have been. How "uppermost" in our minds is the "question of belief or disbelief"? Do we only seldom consider the rationality of our beliefs? Are we victims of our own form of middle-aged complacency (perhaps long before or after middle-age)? How conscious is our faith? Are there explorers inside us?

What makes some people explorers and others not? I have been promoting the idea that exploration is essential to the construction of coherent faith but a jealous and overly zealous mechanic can stifle a malnourished explorer and render him impotent, if he survives at all. Paradoxically, all mechanics start as explorers. But, in order for the explorer in us to develop and thrive, we have to recognize the danger of making our mechanic temperament preeminent. In fact, we are headed in the right direction when we recognize that the mechanic exists to support the explorer and that the competition between maintenance and progress should always favor progress.[10]

9. Lewis, *Discarded Image*, 181.
10. Despite the academic disagreement over the concept of progress (cf. Yerxa, *History of Science and Religion*), I am confident that most people have both an intuitive understanding of it as well as an expectation for it, particularly with respect to their

Ultimately, people become explorers when they recognize the value of doing so. As Josh Billings puts it, "If there was nothing but truth in this world, a fool would stand just as good a chance as a wise man".[11] Perceived value drives all of our decisions and as long as we see greater value in the status quo we will have no incentive to invest in exploratory ventures. As I noted above, however, only the explorer in us is capable of rationally assessing value. This presents a classic conundrum where such an appraisal is possible only if one already has an explorer's disposition and where without it one cannot see the value of having it. Fortunately, in such circumstances we may still become explorers under duress. Crises of faith can engender unquestioning acceptance of the current situation or lead us to abandon a belief altogether but they can also turn us into explorers. This is when we thank the mechanic in us for identifying the problem but are inclined to suspect that the usual maintenance procedures may no longer suffice. That realization can be the starting point but it is feedback from the subsequent probing and discovery that will fuel or deplete our energies as explorers and make us more or less inclined toward future exploratory ventures.

This doesn't mean that every wife should tail her husband just because he comes home late a few times but that, if her faith in his fidelity is sufficiently shaken to do so, whatever evidence she discovers will help prevent that faith from dissolving into wishful thinking. When one fully appreciates the need for evidence the value of exploration becomes apparent as the primary means of obtaining it. Once we have become convinced that the efficacy of a political agenda, the merits of a religious claim, or the status of environmental change can be best addressed by considering the evidence, we will have the necessary impetus to adopt the explorer's perspective and begin a quest to acquire that evidence.

Facing Our Fears

No one will become an explorer as long as fear has its way. When that happens, the resulting phobia—fear on steroids—can undermine the goal of building a rational faith and cripple any progress toward it. Fear is not without merit, yet if we permit it to rule we will leave the exploring to others. Exploration is not for the faint of heart.

---

own lives (cf. McClay, "Revisiting the Idea of Progress").

11. Billings, *His Sayings*, 92. He might have added that in the presence of nothing but truth, exploration would have no value and one could safely remain a mechanic forever. There would be no such thing as faith and hanging onto one thing would be just as good as hanging onto any other.

Fear of the Unknown

No one will become an explorer as long as fear of what may be found overwhelms a desire to know the truth. The wife, political analyst, or religious devotee who prefers ignorance to upheaval or whose grasp on hope precludes their reaching for knowledge are not apt to become explorers. It may seem ironic that anyone who entertains the idea of exploration could be subject to this fear. What, after all, is the point of exploration if not discovery? But any serious attempt to refine the validity of one's beliefs means relinquishing control over the direction those beliefs might take—a scary but necessary prospect if there is to be any hope for an honest quest.

Much of our comfort is derived from familiarity, and the thought that a fundamental belief might be uprooted to be replaced with who knows what might be all the reason needed to shun exploration.[12] But ultimately it is the lure of the unknown that drives exploration. Therefore, by suppressing this fear long enough to get started, the explorer can expect to see it diminish in proportion to his or her discoveries. Explorers know that, no matter how frightening the prospects, searching for an unknown truth is preferable to sitting on a possible lie.

Fear of Slippery Slopes

A more insidious fear but one related to fear of the unknown is fear of slippery slopes. In the arena of faith, this is the idea that if we move even slightly from a current belief, we could easily end up falling away entirely. Alternatively, if we give a competing belief an inch, it will take a mile. This is actually a dangerous fear, not only because it is so pervasive but because, in its extreme forms, it refuses to consider even the slightest contrary evidence. In fact, those held prisoner by this fear are often unlikely to entertain the thought that they might be mistaken. The inevitable result is that even a small step toward modification of their beliefs is impossible. This is not their concern, of course, since their beliefs are infallible and there is no need to improve them.

Ironically, fear of the slippery slope can really be fear of the known—not the fear that we will end up replacing a treasured belief with some unknown commodity but with one that we currently find unattractive. This unbecoming fear is easy to see. Consider:

---

12. Brian Boyd notes that, "Even for those with training, looking for potential refutations of cherished ideas is both emotionally difficult and imaginatively draining" ("Purpose-Driven Life," 33).

- The Democrat or Republican who refuses to believe that there could be anything of value in views of the opposing party for fear that doing so might necessitate voting a split ballot.
- The spouse who refuses to acknowledge flaws in his or her relationship, even though it offers the chance to strengthen the marriage, because of fear that it could precipitate a divorce.
- The young-earth creationist who believes that accepting an evolutionary origin of species places all of his religious beliefs in jeopardy.
- The sociobiologist who thinks that the scientific edifice will be undermined if religious beliefs are permitted any foothold.

In all these cases and others like them, the fear that a small (mis)step could result in a catastrophic slide into an undesirable set of beliefs excludes competing evidence without a hearing. The person at the heart of this regrettable state of affairs is all the more resolute because of reliance on the current belief as the basis for determining what is and is not desirable. Otherwise, why the presumption of catastrophe? In such an attitude it is hard to miss the implicit acknowledgement that the current belief cannot stand up to scrutiny.

We need not pretend that there is no dilemma here. Whether it is worse to hang onto a false belief or risk losing one's faith in something that happens to be true is a genuine concern. A small disturbance can sometimes generate a massive avalanche and with great loss. The mistake, however, is to see this as a foregone conclusion or to neglect the fact that such a cataclysm may be necessary to precipitate greater stability. Yet that is precisely what those with the slippery slope outlook do. Dramatic and costly loss of faith is possible but not inevitable. Explorers are building competence to validate beliefs in a variety of illuminating ways including the ability to take a side trip here and there without necessarily altering the overall direction of their journey.

Moreover, if it is true that, "Humans cling to beliefs ferociously, because they are a core part of our identities,"[13] one way to deal with this fear is to rethink our identities—to see ourselves as dynamic rather than static; to see growth and development as central to who we are; to add to the other beliefs that define us the idea that truth is worth pursuing at all costs. Recognizing that it takes more energy to hold on to a weak belief than it does to let go is a good start. It is far easier to maintain a rational belief than an irrational one and one in which we are secure than one about which we have major doubts. Perhaps we hold so tightly to some just because they are

---

13. Mooney and Kirshenbaum, "Why America is Flunking Science."

harder to hold. In any case, it is impossible to let go of one belief without grabbing another; that is not an option.

Fear of Effort

Unfortunately, just anticipating the exertion required can scare away a would-be explorer. Hard work, particularly when it is in addition to whatever else we are doing, is not quickly embraced and the very thought of spending significant energy and time to read, think, question, and discuss while sacrificing other pursuits in the process is frightening. Without proper perspective, one may abandon the search before it is even begun.

Ah, perspective! What doesn't it color? Is the task before us an obligation or an opportunity? Does it portend instability or illumination? Perceiving that a venture will be titillating or tedious, exasperating or exhilarating doesn't merely influence what the explorer finds—it is also crucial to the decision whether to embark at all. When fear of the impending effort looms large, how one views comfort can make the difference in going forth or staying put.

Most mornings, for example, many of us are faced with the same dilemma: to drag ourselves out of bed long before it is time to get ready for work or school in order to run or cycle, or to turn over and go back to sleep. During much of the year it may be pitch dark at that hour and on occasion quite cold. At other times the heat and humidity can be oppressive. The question is one of effort and the least effortful activity is plain—or so it seems. Through the years, however, we've learned that the comfort associated with the knowledge that we are doing something which is good for our health and that will equip us to climb, kayak, ski, or just keep up with the kids is more than a match for the comfort of falling back to sleep.

Similarly, although it is comfortable to make evidence fit existing beliefs and even easier to accept something without evidence, isn't it significantly more comfortable to know that we have actually searched diligently for evidence and allowed it to speak for itself? Isn't it more comfortable to hold a rational belief acquired with effort than one whose rationality is unknown because it was deemed too uncomfortable to inquire?[14] Explorers are more afraid of stagnation than effort and of having a meaningless faith and not knowing it than of finding it meaningless. Like Mallory, they know that "To refuse the adventure is to run the risk of drying up like a pea in its shell."[15]

14. All this is reminiscent of C. S. Lewis's perspectives on whether Christianity is hard or easy (*Mere Christianity*) but his observations generalize to any domain. I think Lewis would agree that comfort is relative.

15. This quote from Mallory reputedly occurred during one of his public addresses

You may have noticed that the more productive perspectives on comfort come when we are motivated by factors higher in Maslow's hierarchy than the usually dominating physiological drives that constantly stare us in the face. Such higher motives inspire any determined drive for success—a compulsion that can also subdue the fear of effort. Olympians train furiously with a gold medal in mind. Mountaineers brave the elements with expectations of reaching the summit. Countless people suppress their aversion to work and go each day in anticipation of a paycheck. Just the prospect of success is adequate motivation for many worthwhile ventures, despite the effort involved.

Fear of Isolation

People are frequently less fearful of effort if they have company, but what if there is none? What becomes of people who will not exercise unless someone goes with them? Fortunately, no one ever need proceed alone. Every time we adopt the explorer's mien, we enter into a vicarious bond with all manner of explorers, past, present and future, and that bond can help us confront the efforts of exploration in the knowledge that we are in good company. Our struggles with religious faith are a voyage with Job, Aquinas, Augustine, and Mother Teresa. When we ponder the current state of scientific understanding we are looking at the same universe as Galileo and Newton and Einstein. Assessment of social and political mores occurs in the company of Jefferson, Lincoln, Anthony, Gandhi, and King. Although it may be reassuring to think that we might find someone with whom to directly share our quests of faith, it is unlikely their credentials will be so impressive.

Fear of Failure

By this time the pessimist will be thinking of all those explorers I have not yet mentioned—the ones who were shipwrecked, eaten by cannibals, returned empty handed, or embarked on what history has since deemed a senseless pursuit. But some failure is inevitable in a life of exploration and we only know of those individuals because of their efforts. Even failure is a matter of perspective. Beliefs in cities of gold and fountains of youth can only be refined via exploration, and the failure to uphold one belief is always the successful promotion of another.

(Jenkins, "Good Company of the Dead").

As in most investment scenarios, there is greater inclination to expend the necessary effort if success seems assured, and the less confident we are of victory, the harder it is to commit in the first place. When it comes to obtaining a rational faith, the issues are especially fuzzy because the specific details of the quest are only defined by the quest itself. In other words, we cannot assume ahead of time that we will arrive at some particular conclusion. To do so is to delude ourselves about being explorers. In this case it is the pursuit of rationality—the ability to give a reason for our beliefs—that serves as the primary motivator. Not trying, of course, is guaranteed failure.

Fear of Cynicism

We might label the fears discussed so far as the embarrassing ones, and yielding to them is not to our credit. This fear, however, has a more honorable pedigree. Concern that a perpetually inquiring demeanor will lead to a deep-seated and permanent cynicism is a sign that we are conscious of its perils.

To understand this, it is important not confuse fear of cynicism with fear of skepticism. Although you may find your thesaurus equating them, that is too general a comparison and we can use these terms to make a distinction between two approaches to belief that differ in subtle but important ways. The skeptic looks at a subject without automatically accepting its claims to truth but with an eye for evidence and the possibility of being convinced by it and then moving on. The cynic looks at a subject not in the hope of finding truth but of finding something to criticize. Skeptics want to know; cynics want to complain. Explorers must be skeptical: Is that scouting report correct? Are those overtures genuine or a ruse? Is this the best time to move forward or should we delay? A cynical explorer, on the other hand, will find something wrong with any alternative. He may move this way or that but he will do so in a world of grumbles and what-ifs.

Because he is looking for something besides truth, the cynic's attempts at exploration will bear little fruit. Whereas the skeptic filters things with a mesh fine enough to capture morsels of truth, the cynic's net is full of holes so large that nothing sticks. The skeptic walks through a jungle wary of what might want to eat him. The cynic is a skeptic with paranoia—everything wants to eat him. Skepticism is a calling, cynicism a disease (or at best a quickly tiresome attempt at humor). A healthy dose of skepticism sharpens our senses and makes us pay more attention. It is hard to imagine what constitutes a healthy dose of cynicism. Skeptics make interesting partners in a search for truth. Cynics quickly wear out their welcome and with good reason because they offer nothing to replace the objects of their disparagement. It is true that skepticism can degenerate into cynicism, but only if we let it.[16]

The Making of an Explorer

Let's pretend that it's your birthday and I give you an elaborately wrapped present, telling you that it contains my book. Because you are reading this now, you have no need for another copy and, therefore, do not open the gift. In fact, you put the package on a shelf in your closet and promptly forget about it. Several times during the year someone asks you what I gave you for your birthday and you answer with a detailed description of the first seven and one-half chapters of this book because that is where you stopped reading. All the while, my gift to you remains unopened.

So far I have been promoting the idea that exploration is how we obtain a rational set of beliefs but it is not just one of the means, it is the only means. If we consider the vast number of things we believe, it is apparent to all but the most conceited of us that only some of those beliefs will be right. Of the ones that are, there is scant justification in labeling them rational if they were simply accepted as a gift. Although the source of beliefs that have been passed down will give us more or less assurance of their validity, until we actually do a bit of investigating ourselves, our faith may be lucky but that doesn't make it rational. Credibility of the source, in fact, is itself one of those beliefs. Even if you know me, for instance, accepting at face value what I told you might be a bit risky. Perhaps I am an inveterate liar, a prankster, or just trying to make a point. In this case, as in many others, belief about the contents of a message is tantamount to belief about the credibility of the messenger.

Hopefully, my example never made it to completion in your case because you rushed to your closet to open the present after reading the previous paragraph. Regardless, once you do eventually check the contents, the story can unfold along several lines: there is indeed a copy of this book in the box; it is an autographed copy and hence very valuable (at some mythical future time); it is a book I wrote but not this one; it is a book belonging to me but not one I wrote; it is a box of chocolate covered cherries which have become quite stale; it is . . . well, you get the picture. Yet only one of these scenarios corresponds to your original belief (whatever it was) about the

---

16. I can hear objections out there from all of the Menckens and Bierces. All of us have our cynical moments but it is cynicism as a habit that I am censuring because it offers nothing productive to the pursuit of rational faith. Call this my cynical moment if you wish.

contents of the gift and there are more ways for you to be wrong than right. In short, examination is critical to the rationality of your belief.

You might protest that, once the gift is unwrapped, faith no longer plays a role because then you would "know." But that is not quite right. All you have really done is to accumulate some evidence that what I told you was true. To further buttress your belief in my honesty, you would also need to verify that I actually did write the book, that I wrote all of it, that it really is my gift and not one I was to deliver for someone else, that it was truly intended for you, and so forth. That probably seems like more effort than is warranted to verify the truth of a statement so simple as my original claim but this is, after all, just an illustration. The key is to decide when a belief is important enough to merit such deep probing. Every time we face a challenge to our beliefs we are handed an invitation to lead our own expedition. We can't (and shouldn't) respond to all of them but it is the habitual explorer who is better able to make that call than a weekend voyager.

But what is the origin of the person for whom exploration is a habit? Was Columbus born with wanderlust in his eyes? Were his motivation and tenacity and skill a cradle gift most of us somehow missed? The natural inquisitiveness of infants may indicate a genetic predisposition to explore but the type of exploration needed to undergird our deepest and most important beliefs demand a level of commitment and sustained effort far exceeding those humble beginnings. Yet we are all born with this potential. Every day we have the option of accepting without question what we are given or of venturing beyond. Like the decision to exercise regularly, however, such choices are ultimately about the lifestyle we intend to embrace. Day-to-day activities then follow that lead. Hoping that we will just drift into the habits of the explorer is akin to hoping we will drift into marriage or a job—it may happen on occasion but the prospects for doing so are quite poor.

Vision

Food, weaponry, leadership, and negotiation skills probably proved essential to the successful completion of many exploratory ventures, but I suspect that if we could poll a collection of famous explorers they would say that understanding their mission and appreciating its significance were more important than any of these. No one becomes or remains an explorer without vision—a synergistic blend of hindsight, foresight, and insight that recognizes valuable possibilities and how to achieve them. Whether exploring the northwest or the North Pole, investigating mountains, moons, or molecules, or circumnavigating the globe by sailing vessel or spaceship, vision is perhaps the most essential characteristic any explorer can possess. No genuine explorer lacks this quality and no exploratory venture of any consequence can be conducted in its absence. From believing a quest is worthwhile in the first place to continued progress in the face of fatigue and discouragement, vision is indispensable.

For most explorers, their actual sight is quite limited. After all, if they knew everything they would find there could be no exploration to begin with. It is probably safe to say that Meriwether Lewis and William Clark never envisioned a Gateway Arch or millions of web pages paying them homage but it seems an equally safe bet that their vision was every bit as expansive. For those exploring the rational foundation of their beliefs, vision is no less important. Such exploration is the gateway to truth and even a metaphorical arch symbolizing the possibilities and potential consequences envisioned should be incentive enough.

But this is just one of the ways that explorers see matters of belief differently than do mechanics. In general, explorers are geared to see things from many points of view. The French writer Proust suggested that,

> The only true voyage of discovery, the only really rejuvenating experience, would be not to visit strange lands but to possess other eyes, to see the universe through the eyes of another, of a hundred others, to see the hundred universes that each of them sees, that each of them is.[17]

Even if there is another "true voyage of discovery," the ability and willingness to see deeply into the heart of a matter—to broaden one's perspective—is of the utmost importance if one is to make the best decisions and avoid appearing, in retrospect, narrow-minded, shortsighted, and perhaps even silly. Consider, for example, the observation made by Joseph Ives, leader of an 1857 expedition that reached the confluence of Diamond Creek with the Colorado River in the Grand Canyon:

> The region is, of course, altogether valueless. It can be approached only from the south, and after entering it there is nothing to do but leave. Ours has been the first, and will doubtless be the last, party of whites to visit this profitless locality.[18]

Anyone who has ever been to that bleak area can understand what he was thinking. Ironically, Powell was there a scant twelve years later,

---

17. Proust, *In Search of Lost Time*, 343. There is no need to assume those views are equally useful but one view can sometimes put another in relief.

18. Crumbo, *River Runner's Guide*, 53.

approaching from the other direction,[19] and thousands of people each year now end their river trips through the canyon at that very location. Ives may be forgiven for only seeing so far—explorers aren't infallible—but if he had tried to look at Diamond Creek the way others looked at Manhattan Island when it was still forested; if he had recalled the way Michelangelo looked at stone or Shakespeare looked at words, he might have been less hasty to pass judgment that turned out to be primarily about his own shortsightedness.

The explorer is constantly wondering: "What am I missing? Is there another way to look at this? What do others see that I don't? Is what they (or I) see really there? Am I so focused on one thing that I miss something of far more significance?" Ebenezer Bryce reputedly described a convoluted piece of arid real estate in southern Utah as a, "hell of a place to lose a cow,"[20] but losing one there today would be difficult because any number of tourists from Japan would probably find it in a matter of minutes as they admired the grandeur of Bryce Canyon National Park. Bryce, it turns out, was lucky, for seldom will those without "other eyes" have a park named for them or otherwise be honored. Why should they?

Regrettably, there is the very real possibility that you will see something that doesn't exist. Every year, countless failed businesses offer mute testimony to the fact that people saw potential that wasn't there. But explorers accept the risk of looking foolish. Those who believe that the truth will set them free[21] (and see the alternative as something like a prison) will make its pursuit an enduring priority. This, too, is a function of "other eyes."

Attitude

The extent of an explorer's vision influences subsequent attitudes and behaviors. Although all of the inevitable twists and turns are not foreseeable, belief that the search will be illuminated by a pervasive light makes it possible to progress with confidence in fruitful directions. John Bunyan had a special appreciation for this:

"Whither must I fly?"

"Do you see yonder Wicket-gate?"

"No."

19. Ibid.
20. Krell, *National Parks*, 139.
21. Religious perspectives on faith and truth are considered in a subsequent chapter but Pontius Pilate's cynicism about human inability to define truth (John 18:38) can be contrasted with Jesus' claim that "the truth will set you free" (John 8:32).

"Do you see yonder shining light?"

"I think I do."

"Keep that light in your eye, and go up directly thereto: so shalt thou see the gate . . ."[22]

The explorer doesn't consider it obligatory to presume what is behind a gate, although there is an intense desire to know. Because the main priority is to find and open the way to truth, success is not gauged by the confirmation or refutation of an existing belief but by the level of rational support accumulated at each entry point.[23] Confidence that such support can be amassed is not some idle hope but is based on observation of countless successes by other explorers. That confidence fuels the explorer's faith in what is possible and is reflected in the attitude taken toward temporary setbacks or an apparent impasse. This is not to say that explorers don't start with large measures of hope. But they also start believing that, as they journey, hope will be replaced by a growing understanding moving them closer to the truth. If you were wondering how much faith is enough to be an explorer, that is how much.

Explorers are also not troubled by the fact that they may not be first on a scene. Lewis and Clark "discovered" things Sacajawea knew intimately; Columbus's new world was another's old home; students in every generation discern principles already known to their teachers and their teachers' teachers. Even in science we find evidence of rediscovery or parallel innovations.[24] Particularly when it comes to beliefs, someone else's discoveries are hardly enough to stifle a search although they may have value as evidence. In fact, no matter how many people have previously trod the same ground (including parents and one's most influential teachers), the rational components of the faith they have acquired are only indirectly transferable. Because beliefs passed on include functional as well as faulty ones, scrutiny is required by each new generation. That was the thrust of the birthday-gift illustration—unless we bring something to a belief, we can't claim it is rational.[25]

The bottom line is that rational faith requires more than rubber-stamping the beliefs of others. In that sense, every belief is virgin territory. Unfortunately, a consumer mentality can spill over into the arena of belief such that we permit or even expect others to do the analysis for us. Although

---

22. Bunyan, *Pilgrim's Progress*, 7–8.

23. Recall that prescribing a specific result prior to a search is not exploration but rationalization.

24. Cf. Simonton, *Creativity in Science*.

25. Compare Polanyi (*Personal Knowledge*) on "commitment."

there will always be matters in which we lack the training or skills to explore at a high level, adopting an attitude of surrender guarantees that we will never achieve more than a parrot-like faith.

We don't have to look far to see how this works. Blindly accepting what some reviewer believes constitutes a good movie or book is a sure way to miss many that are enjoyable (or to waste time on many that are not). I would have never climbed a mountain if I had shared my parents' belief about the value of doing so. You can probably think of comparable examples, but probing the beliefs of others is especially consequential in areas where things aren't as simple as movies and hobbies. A parent, for instance, may deeply desire that her child follow her belief footsteps in matters of religion, but of what credit is it to the child who does so if the sole justification is, "This is what mom believes"?[26] Recall that it was only Galileo's willingness to go back over the paths already traveled by Aristotle that enabled him to see where Aristotle had been off course. In short, explorers will no more think of letting others believe for them than they would permit them to climb a mountain, go to college, exercise, or travel to Europe for them.

Behavior

Whether trying to find a new outfit or seeking a cure for cancer, humans are constantly searching for things. When you think about it, you will find that almost everything you do can be classified as a type of search (That thought process itself will involve searching through a list of things you do to see if they qualify). We begin the day looking for something to wear and end it trying to find an entertaining television show to watch or book to read. At work or school we seek solutions to problems and several times a day we are searching for food to prepare or a place to eat. Once or twice a year we look for a suitable place to spend a vacation and for some period of our lives many of us are engaged in the search for a mate. Throughout this book I have been searching for the proper things to say as well as how to say them.

It is probably apparent from the list above that many of our searches are important primarily to us and, frequently, only in a temporary way. Other inquiries, however, may have significant and lasting impact. It is of no consequence to you or anyone else if my search for something to eat results in Mexican or Chinese cuisine (or just a peanut butter sandwich) or even if I miss lunch altogether, but my search for a spouse had implications for

---

26. The operative word is "sole." A child whose parents are credible in other ways will have reason for trusting them with respect to matters such as religion but that cannot *make* those beliefs correct.

numerous people including at least several not yet then born. On the other hand, notable geographic, scientific, philosophical, and religious quests such as those of Columbus, Newton, Socrates, or Augustine have influenced untold numbers of individuals and it is likely they will continue to do so for ages.

Our investigations thus occur on a sliding scale of importance, populated by beliefs about many things of little consequence but by a few with the potential to affect our physical, mental, and spiritual well-being, define our sense of meaning and purpose, determine the kind of lives we lead, and impact our influence on others. Figure 1 provides a rough visual idea of the relationship between the importance and number of one's beliefs.

Figure 1. A plausible relationship between how many things one believes and the relative importance of those beliefs

Although most searches are mostly automatic—dare we say, mechanical—genuine explorers direct attention to what they believe to be major issues. Their intentionality may filter down to topics of less importance, but they can never escape the attraction of what really matters.

What is it, then, that the explorer of beliefs is so intent upon finding? Explorers want truth but search for evidence in the belief that, short of a revelation,[27] evidence is the only avenue to truth. But they aren't relying on an oracle to tell them if one theory, job offer, or potential spouse is superior to another. In fact, explorers see evidence as a kind of revelation but they

---

27. I began to address this thorny issue in a previous chapter and will have more to say about it later.

don't wait for it to come to them. They actively search for it, evaluating, collecting, discarding, and weighing alternatives to support or refute whatever belief is currently on the table.

Not everyone agrees. Simone Weil, for instance, claimed that, "We do not obtain the most precious gifts by going in search of them but by waiting for them. Man cannot discover them by his own powers and if he sets out to seek for them he will find in their place counterfeits of which he will be unable to discern the falsity."[28] I think it is easy to appreciate the thinking behind such a statement—it places a premium on the presumed beneficence of a higher being and rightly denies our own omniscience. But is it an adequate position from which to construct a sound faith? Is it true that exploration is not only unnecessary but actually harmful? Those who wait for a spouse, health, and education to come to them are likely to end up lonely, decrepit, and ignorant. We may argue about what constitutes "the most precious gifts" but it is hard to contend that accurate beliefs are not among them. Will truth simply come to those who wait? And what does it mean to wait? Truth may occasionally fall into idle laps but, in the absence of investigation, on what basis should it be validated?[29]

The foundation for productive search—knowing where to look, what to look for, and how to assess it—is active learning. That is how to acquire "other eyes." In one sense, of course, you can't help but learn. For example, you can probably describe all kinds of products that you have seen advertised on TV that you would be completely happy not knowing about. However, the marvelous sponge that is your brain has absorbed their descriptions in a non-intentional act of learning. Passive learning is easy but makes us subject to manipulation. Active learning, because it is intentional and requires thinking about what is being learned, can be difficult but puts some measure of control in our hands.[30]

Consider, for example, the following excerpt from a news article entitled "Mountain Climbing Bad for the Brain":

---

28. Weil, *Waiting on God*, 56.

29. There is no need to deny religious convictions that there are things we are incapable of doing for ourselves. But Christians are encouraged to be active inquirers (e.g., Matt 7:7; 13:45–46) and orthodox Jews and Muslims have never waited for Jerusalem or Mecca to come to them. Biblical passages such as Ps 27:14 and Isa 30:18 which seem to extol the benefits of waiting on God might better be seen as commending patience but, either way, they only make sense if it is believed that there is evidence to support their claims. One will only wait for a God who is deemed capable of justifying the wait.

30. Inductive reasoning typically operates without much conscious reflection (i.e., easily) whereas deductive reasoning seems to require considerably more effort.

> Six of the nine climbers had lower than average scores on the Digit Symbol test. . . . Three out of nine scored lower than average on memory tests, while four scored below average on a visual-motor function test.[31]

A loose reading (focused on the "lower than average" or "below average" labels for all tests) might seem to support the contention that cognitive deficits can be induced by high-altitude mountaineering. A slightly deeper engagement, however, should be sufficient to make one wonder if, in contradiction to the intended message, it would be prudent to become a mountaineer in order to improve cognitive ability.[32] Of course moving past raw information to a deeper understanding will, in many situations, require considerably greater investments of mental energy than needed for our simple example.[33] Such intentional thinking can shake one's faith in what would otherwise be a shallow conclusion but that is just what the explorer wants.

Ironically, a shallow form of passive learning often takes place where we might least expect it: educational institutions. There are several contributing factors. For one thing, much of the learning that occurs in any school is coerced. Most institutions have a set of required courses that have been deemed important to the student's ability to understand the world and think critically but, unless students see the value in such courses, their primary objective will simply be to pass the exams. In such cases long-term learning is imperiled, becoming an unlikely byproduct rather than a useful foundation. It is also easy for students to succumb to passive learning because the added rigors required for an active approach are time-consuming. Even with a genuine desire to dig deeply into a subject, time constraints can prevent more than a cursory examination. Furthermore, the perceived authority of a professor or their own rebellious tendencies may lead students to consume or reject whatever is served without question. In both cases learning can stop short of the critical inquiry needed to make it truly useful. But if passive learning can occur in the very setting where people are hoping to prevent it, how much greater might be its threat elsewhere?

Active learning, then, is the fuel on which reason runs. Remove it and the rational engine stops, although the imaginative one can continue to chug

---

31. Parker-Pope, "Mountain Climbing Bad for the Brain."

32. Hopefully you noticed that on two of the three tests the climbers examined actually scored at or above average!

33. Actually, the example given is not as simple as might first appear. There are a variety of other reasons for wondering if the conclusions are justified and, to be fair, one would need to refer to the original research.

right along creating friendships that don't exist, fabricating alien abductions, harboring irrational fears, and accepting all manner of things found only in dreams. Active learning is essential to keep us in touch with reality.[34]

It's not that folks intend to segregate learning and thinking but that they don't intend not to. One particularly productive strategy for integrating these (and for acquiring those "other eyes" recommended by Proust) involves examination of ideas from other fields of study beyond the one immediately at hand. Interdisciplinary approaches provide metaphors, analogies, and other insights for understanding and analyzing a problem in ways that would otherwise never occur to us. Computer scientists, for instance, employ techniques for solving optimization problems based on principles observed in biology and metallurgy.[35] Psychology provides new ways for individuals to think about the words and deeds of Shakespearian characters. Anthropology and sociology shed light on various religious issues (and vice versa). Some of these examples may look as if they are far removed from the world of faith but they are not. Trying to leverage old insights in one area into new insights in another is a meaningful way to assess existing beliefs and to improve their rational foundation.

Individuals whose behavior is characterized by deep curiosity, active learning, and creative thinking never consider themselves done. They are aware, with John Stuart Mill, that, "The fatal tendency of mankind to leave off thinking about a thing when it is no longer doubtful, is the cause of half their errors."[36] They also know that inquiry can seldom stop with one pass because there is usually nothing to tell them when they are done.[37]

On the surface, this appears to present a hopeless dilemma—the goal is unassailable faith but obtaining it is impossible. Despair is an option but not an attractive one. Therefore, before you shoot yourself, we'll consider a productive alternative. First, I need to make it clear that this will be most meaningful for those who feel at least some concern about their inability to reach the heights to which they aspire. Complacency suggests that the dilemma has not been recognized, that there is really no aspiration to build

---

34. The caricature for trying to think without learning is the armchair quarterback who, despite having no access to scouting reports, having seen no practices, knowing no players personally, lacking knowledge about their current physical condition, and in most cases having never played college or professional football, nonetheless considers himself competent to call the plays. Unfortunately, a fair amount of this same sort of thing goes on in more serious arenas.

35. Techniques termed "genetic algorithms" and "simulated annealing," respectively.

36. Mill, *On Liberty*, 41.

37. This does not mean there is not *someone* who is ready to tell us when we are done.

a rational faith after all, or that it is thought one has already arrived. None of these attitudes are conducive to developing meaningful faith but they provide much better reasons for shooting yourself than does recognizing the impossibility of such a quest.

The way to handle this quandary is to understand that the reason for making perfection a goal in the first place is to keep us moving in a positive direction. The alternatives are stagnation or, worse, regression. If any old spouse, job, house, and so forth will do, that is probably what we will end up with. Shoot for the stars and you will at least glimpse starlight.[38] The result of shooting for impregnable truth—for rational faith that can sustain any assault—is an army of beliefs that can survive and thrive among a host of less fit competitors.

Judgment

In the quest to optimize beliefs, as with any optimization problem, there will be better and worse solutions. Although the goal is to find the very best, the search is typically conducted by exploring those solutions currently thought to be so. Metaphorically, we may never summit the highest mountain, yet as long as our view is occluded by taller peaks we will keep climbing.

But, even if we were to reach the pinnacle of truth, how could we know it? How could we justify calling a belief a fact with the confidence that it is undisputable? If you've been paying attention, you know that the answer is, quite simply, "We could not." In other words, our inability to know that a belief is unassailable is one reason it cannot be (even if it otherwise is). Nevertheless, though we may struggle to know that we have reached the truth, we still need some assurance that we are progressing toward it.

Ideally, such confidence comes from being able to discriminate between defensible and indefensible positions. Deciding one way or the other is also a belief, but what alternative is there? Unfortunately, the all-too-common tendency to staunchly defend poorly grounded beliefs makes it appear that this standard is worthless. Besides, doesn't the defensibility criteria confront us with the same dilemma, merely pushed back one level (i.e., defending the belief that another belief is defensible)?

These are valid concerns but genuine explorers recognize as much and then perform an honest assessment to determine whether they are truly engaged in a relentless search for evidence and a legitimate attempt to

---

38. Note the parallel to Jesus' injunction to, "Be perfect, therefore, as your heavenly Father is perfect" (Matt 5:48). As for any seemingly impossible expectation, those who take it seriously are more likely to keep moving toward the goal.

assimilate it into a coherent framework (or only kidding themselves). They are thus able to answer a fundamental question about any particular belief: "Does it work?"

Judging how well a belief works can be tricky because superficial views may support some belief that a more thorough investigation will reveal to be mistaken. In other words, a belief deemed suspect at one level of analysis might appear all but invincible from another. For instance, it could be hard to believe that the surface of the earth is subject to massive change if considered from a limited geographic and temporal perspective. The accumulation of evidence, however, provides overwhelming support for dynamic crustal activity and is widely accepted today.[39] Any belief can likewise be modified or supplanted as additional evidence and interpretive insight become available if the new belief generated works better than the old one.

But what does it mean to say that a belief works? Ultimately, we judge that a belief works because it passes the relevant tests to which it is subjected. Deciding what is relevant is also a crucial belief but, even if we disagree about the appropriate tests, we can be certain that if a belief is untested we cannot know if it works. Not knowing that means its rationality is tenuous.[40]

After injuring my leg during a long run when I was an undergraduate, a nutritionist prescribed massive doses of vitamin C and bioflavonoids as a means of enhancing the ability of the muscles and tendons to take the abuse to which I was subjecting them. With a mixture of belief and hope that I would incur fewer injuries, I purchased the recommended product and proceeded for months to pop several pills a day.

Now, the question: Did my belief work? (Did his?) Although I repeatedly conducted one test (taking the pills), I cannot say. I don't recall any subsequent leg injuries but that may have been the result of running less. Neither do I recollect having many colds during that time but that has never been much of a problem anyway. The only definitive change I noticed was a discoloration in my urine—a predictable result when one is eating the equivalent of eight and one-half oranges a day. On the other hand, I do know that I began to lose my hair around my sophomore year in college, which was close to the time I was supposed to be doing my body a favor by taking those pills. Should I

---

39. Cf. Jastrow and Rampino, *Origins of Life*.

40. Bacon recognized that testing was essential to building a solid foundation for the sciences but, as with his other suggestions for improving scientific understanding (e.g., looking for evidence, testing even the basics, not relying on logic alone, being aware of what can influence decisions, perceiving fallibility of the senses, and changing the level at which things are evaluated), this can also be useful for analyzing other beliefs (Bacon, *Great Instauration*).

change my belief in the efficacy of vitamin C for muscle protection to a belief that it contributes to hair loss?[41] Are the beliefs incompatible?

Suffice it to say, these are not life's big questions and I have moved on, willing to live in belief limbo in that regard rather than spend the time trying to refute one hypothesis or the other. Yet if I believed that knowing the truth about those pills was germane to my fitness or if I had been a movie star (for whom hair loss could spell contract trouble), I might have been more inclined to pursue the matter further. What did the actual results of the nutritionist's work with other athletes show? Were they well-controlled studies published in peer-reviewed journals? Have others been able to replicate his results? Were there genetic factors that made me more or less susceptible to the effects of the vitamins or that contributed to hair loss? Accumulation of such evidence would be the route to deciding if the original belief "worked."

Such evaluations sometimes take on a see-saw persona where one piece of evidence appears to support a belief while the next seems to refute it. In such a scenario, it is possible to adopt or discard the belief based on the last piece of evidence but that is seldom a good idea. The very presence of conflict might indicate that more evidence is needed or, more importantly, that the current interpretation of the evidence on hand is questionable. This is not a sign to quit but to keep exploring.

Sooner or later, however, even the restless need a rest. Although all beliefs remain unfinished, one of the practical ramifications of developing a belief that works is that it need not be revisited as often as those for which the support is still highly questionable. One is then freed to put his or her primary energies elsewhere. This is how the explorer—temporarily satisfied but never complacent—manages to live with less than perfection. Exploration occurs on multiple fronts but ebbs and flows with changing perceptions about what beliefs are important and how well they are supported.

Danger! (on the Road to Rational Faith)

As we've seen, the journey to rational faith is beset by numerous obstacles. If that was not so, we could expect to find nothing but well-founded beliefs wherever we turn. The fact that we do not is due to the difficulty in starting

41. This is the issue of distinguishing between causation and correlation (cf. Hume, *Human Understanding*; Anderson, "End of Theory"). Did ingesting massive doses of vitamin C cause my hair loss or was that merely a correlated (but unrelated) event? Was a political candidate's rise in the polls due to a particular comment or entirely coincidental? Did prayer divert a tornado from someone's home (causation) or would it have missed anyway? Does a cereal actively reduce cholesterol or are those eating it less likely to consume a breakfast of eggs and bacon?

such a voyage in the face of misconceptions about faith, a failure to appreciate the need for exploration, or any number of debilitating fears. Even if the quest is begun, wishful thinking, failure to seek the necessary evidence, and a variety of limitations can hinder one's progress. In this section I want to raise one additional warning about an obstruction that can derail even the most intrepid explorer.

The problem is arrogance and it is especially sinister because the danger frequently increases along with one's competency. The same conceit that plagues the (self-identified) invincible athlete, indomitable businessman, or incomparable actress just as surely threatens veteran explorers of faith who are aware of their vast experience, certain that they know what they are doing, sure they can see the problems and pitfalls clearly, and assume they are in control. Arrogance is the result of slipping over the fine line between confidence and over-confidence. The "Look at me, I'm an explorer!" attitude can generate its own form of pride but inexperience and early failures are often enough to bring the neophyte back down to earth. A growing string of successes, however, can result in unwarranted exaggeration of one's prowess.

For example, the more familiar we are with the positive and negative evidence for any particular belief, the easier it is to see both sides. Rather than admitting that the jury is still out, it is simpler to promote an arrogant descent into relativism where one belief becomes as good (or bad) as another. But there is no surer way to end a search than to believe that it is futile.[42]

That is bad enough but accomplished explorers are liable to overrate all of their conclusions. Consequently, although ignorance is bad, arrogance is worse. Ignorance can be followed by a quest to alleviate it but arrogance assumes the quest is over. This cold presumption is doubly insidious because it results in a frozen set of beliefs that represent a mindset exactly opposite to that with which the process was begun. The illusion of rational invincibility accompanying such arrogance may not be the main reason that pride is considered one of the seven deadly sins but it should be. Today we laugh at anyone who believes that the earth is at the center of the universe but there are many people who seem to believe that everything revolves around them (and their beliefs about how God, the universe, etc. works). Is there really that much difference in the two mindsets?

But arrogance doesn't stop there. Once utterly convinced of something, it is a small step toward a tendency to rationalize it at all costs. The one who would rationalize is searching, but the search is characterized by

---

42. Berry suggests that too much knowledge could prevent one from making any decisions at all (*Life is a Miracle*, 149). Shermer's belief that one "cannot prove or disprove God's existence through empirical evidence or rational analysis" is a case in point (*Science of Good & Evil*, 4).

picking, choosing, and manipulating evidence to fit a set of stagnant beliefs. With eyes that don't see and ears that don't hear,[43] the outcome is a neatly made-up mind that is no longer ruffled by thought. Anyone is vulnerable to this hazard, but for the arrogant explorer who thinks he has arrived, it is especially problematic. Despite having all the equipment with which to adjust a beleaguered belief, he no longer opens his toolbox, attempting instead to (mechanically) hold things together with tape and glue. If he is successful at maintaining the façade, he might convince others to accept his point of view. He might even convince himself.

A subtle form of rationalization that can plague expert and novice alike is to think that a consensus view is a source of validation for one's beliefs. When it comes to faith, however, we cannot assume there is safety in numbers.[44] Just because everyone in our fraternity or sorority, university, profession, church, state, country, or world believe something, it might, nonetheless, be false. In fact, no matter what your belief, there will be plenty of people who share it. Is capitalism or socialism the better economic plan? Are Republicans or Democrats more in tune with the needs of the United States? Will Notre Dame or Southern Cal win the college football championship next year? Does Christianity or Hinduism or Islam or Judaism offer the best explanation of deity? You can find just the support you crave for your answer but everyone's answers cannot be correct. This problem is sinister because, even when we recognize it as a possibility, we don't see it happening to us.

Yet as it turns out, there are several reasons why we might try to substitute a consensus opinion for rationality. For one thing, anyone in a democracy such as ours is accustomed to living with sentiments of the majority. But truth is not something created by voting or legislation. That is a good thing, lest truth be constantly changing. Nevertheless, our cultural habits can tempt us to turn rational belief into a popularity contest.

A preference for people who share our beliefs can also lead us to wrap ourselves in a protective cocoon of acquaintances that insulate us from being preyed upon by competing perspectives and in which faulty beliefs can continue to grow and flourish along with good ones. Because of this, once a belief is adopted it tends to be self-perpetuating. Even if we suspect it is in error, pressure from peers can be sufficient to bring us back in line or at least

---

43. Jer 5:21.

44. Compare Hume (*Human Understanding*, section VI) regarding the strength of numerical support as it pertains to belief formation.

encourage us to act the part. In fact, it would be interesting to know in any community of shared belief how many are just acting.[45]

Uncritically adopting a consensus view avoids the hard work of constructing a rational faith but at the risk of becoming nothing more than a reflector of someone else's erroneous beliefs. When others explore for us, we only find what they do. If what they find is wrong, too bad for everyone. Although a variety of limitations can force us to rely on the views of others, without some exploration on our part, how could we possibly know whether their views are credible? The searcher for truth must decide when it is appropriate to play solitaire. Not knowing is dangerous.

Better and Worse Reasons for Believing

In his book *A Devil's Chaplain*, Richard Dawkins includes a letter entitled "Good and Bad Reasons for Believing," in which he warns his daughter and any other readers about the three great evils of faith. According to him, these are tradition, authority, and revelation.[46] I should make it plain from the outset that there are, indeed, reasons to question beliefs formed on the basis of these alone but there is much more to the story than he allows.

In the first place, scientists (of which Dawkins is one) subscribe to tradition and authority on a regular basis in the course of their daily activities—what Kuhn calls "normal science."[47] The key is that scientists are supposed to be prepared to modify their beliefs if circumstances warrant but, as Kuhn has shown us, that is more easily said than done.

It is also hard not to miss the fact that, despite Dawkins's desire to convince his daughter that she should shun those who say, "You must take my word for it," he is asking her to do just that. Should she? Should we? Is the motto of the Royal Society, "Nullius in Verba" ("On the words of no man") to be taken at face value? Isn't there a sense in which it refutes itself? Clearly both Dawkins and the Royal Society are really expressing the idea that testing and verification are essential prerequisites to the acceptance of anything conveyed to us,[48] but castigating authority or tradition is too

---

45. Tendencies to rely on numerical support to buttress belief is not restricted to personal contact with peers. Books, magazines, journals, and other media extend the reach of a shared opinion and the Internet, in particular, has provided new ways to follow the crowd. Cf. Cerf, "Google's View," 120 regarding the problem of accuracy.

46. Dawkins, *Devil's Chaplain*. He missed a great opportunity to capitalize on the acronym T.A.R. for these supposed sticking points in rationality.

47. Kuhn, *Structure*.

48. This seems to be the spirit in which Thomas Huxley wrote, "The improver of natural knowledge absolutely refuses to acknowledge authority, as such. For him,

simple. The fabric of our individual lives and that of society itself hinges in large part on our unavoidable reliance on received knowledge—as we saw earlier, it is physically and temporally impossible to learn everything from scratch.[49] Questioning authority and tradition is one thing; denying, ignoring, or minimizing their contribution is another. The explorer is called upon to make rational decisions about who and what to question and accept but those very decisions will be based on a set of beliefs inevitably influenced by authority and tradition no matter what attempts are made to avoid them.[50]

Sometimes there is a tendency to buck both of these as a way of flaunting one's independence, or to naïvely assume that we are smarter or that our insights are more monumental than those of our ancestors—an attitude C. S. Lewis termed "chronological snobbery."[51] It is worse to believe that we have the last word. Even if some of our ideas achieve the status of tradition and authority for future generations (and we should be so lucky), why should something also subject to change be considered sacred?

Like authority and tradition, revelation is more nuanced than one would gather from Dawkins's treatment. This is not surprising since his intention is to encumber the concept with a debilitating amount of theological baggage, but that neglects the other sources of revelation for which we also must account. In fact, we are typically quite comfortable with the basic idea: she revealed her deepest secrets; the experiment revealed the theory was flawed; the senator revealed his previous affairs; her actions revealed ulterior motives. In essence, revelation suggests that someone or something bestowed knowledge upon another and even a Dawkins—especially Dawkins—should be prepared to concede that, "It was revealed to me by Darwin." As it turns out, besides revealing the basic mechanism of species differentiation, Darwin himself refers to things revealed by Nature, the microscope, geology, and embryology.[52]

In fact, I seriously doubt that many scientists would quarrel too much with the idea of revelation, although most probably prefer to label it "inspiration"[53] or "a flash of insight"[54] and would be inclined to attribute it to

---

skepticism is the highest of duties . . ." ("Improving Natural Knowledge").

49. This is the problem with Dawkins's attempt to skirt the authority issue by assuming that scientific principles that haven't been personally verified can be, in principle. Well, sort of . . . Recall the limitations described in chapter 6.

50. These constitute the paradigms of which Kuhn speaks in *Structure*.

51. Lewis, *Surprised by Joy*.

52. Darwin, *Origin of Species*, 248, 265, 292, 457.

53. As noted in an earlier chapter, Dilday suggests that some people consider the type of inspiration leading to the Bible, works of Shakespeare, etc. as basically the same (*Doctrine of Biblical Authority*).

54. Or "luck," as Einstein described what Walter Isaacson calls his "head-snapping

a fortuitous product of neural activity[55] or something similar. The point is, we all accept revelation as possible, we just argue about the source or want to give it another name. Failing to be skeptical about a supposed source (God, fate, nature, Shakespeare, technical journal, Nobel laureate, neural firing patterns, whatever) is bad but discounting the possibility of a source altogether is worse.[56]

---

insight" into general relativity (*Einstein*, 196).

55. Cf. Hawkins and Blakeslee, *On Intelligence*.

56. Dawkins's sore spot is obviously divine revelation. I deal with that in chapters four and nine.

PART III

Religious Faith

——— CHAPTER 9 ———

The Nature of Religious Belief

The first essential of a working religion is that it shall be congruous with Man; the second that it shall be congruous with Nature. . . . A divorce here would be the catastrophe of reason, and the end of faith.[1]

—HENRY DRUMMOND

The Way We Are (How People See Religion)

WHEN MY OLDER SON Joshua was four and a half years old he woke up one morning after dreaming that he had been scratched by cats. The dream had apparently been fairly traumatic and when I asked him what caused his nightmare he replied, "I think God did." Now Joshua is a bright guy but his theological understanding at that age was understandably incomplete. Although God has been implicated in the dreams of many people, I considered it likely that there was a more proximate explanation for this one. A video, a story, an anecdote from another child, all seemed better candidates to have created the neural context in which the dream developed than the actions of deity. What purpose could God possibly have for making or letting him dream such a thing? After all, malicious cats can hardly rank with life-changing visions of purpose and calling. Of course I am quite ready to admit that God might have had reasons that escaped my notice—I'm pretty sure that has happened on occasion. But I believed it more likely then (and now) that if we were to detect God's involvement in the dream we would

---

1. Drummond, *Ascent of Man*, 32–33.

need to trace the causal chain for some distance before discovering how or why. That strikes me as a decidedly problematic venture because, looking toward its beginning from the end of such an imagined chain, God's involvement in this case seems so far removed as to be imperceptible.[2] However, I suspect that others would disagree. Throughout history there have been those who conceive of deity directly influencing nature and, for them, the causal chain has but one link. For others, the chain is even shorter—there is no causal chain involving deity because there is no deity.

In any event, these are things folks believe about God. They are part of their faith that there is or isn't a God and that, if there is, he (or she or it) has certain attributes, attitudes, and desires and exhibits a particular repertoire of behaviors. Although I have gone to great lengths in this book to make it quite clear that faith is operative in every area of our lives and not just in the religious realm, I am, nonetheless, acutely aware that when most people think of faith the first thing (and, for many, the only thing) that comes to mind is religious faith. Well, here we are. In this section we will be considering religious faith in particular, but it is important to keep in mind that all of the things I have said so far apply to this manifestation of faith no less than it does to others. It is also the case that some of the specific observations we make here can provide additional insights into other applications of faith.

We begin by acknowledging our differences. One person's strong belief in the presence or absence of God is matched somewhere by an equally weak belief in that presence or absence. In fact, there is a side of me that cringes when I hear the word "believer" used to describe a person with a definitive religious pedigree because we are all believers, just in different things and to different degrees.[3] Even in a specific context—a Christian setting, for example—"believer" is too general to be of much use. In some respects it is like the proclamation on U.S. currency: "In God We Trust." Of which God are we speaking and for what are we trusting? Perhaps the mint should stamp "with 0.85 confidence" on one set of coins, "with 0.20 confidence" on another, and so forth, but which would you request? Would it make any difference if you or your loved ones had just been traumatized by or inexplicably spared in a disaster?

The slogan, "In God We Trust," clearly carries certain implications but in one sense it is actually quite safe precisely because it makes no additional

---

2. I am speaking of the causal links in the dream under consideration. In other situations the chain may seem much shorter (or longer). However, the chain metaphor itself is quite possibly flawed since it is merely part of an anthropomorphic approach in the attempt to understand God's role in events.

3. Scott Atran's view of deity—"belief in hope beyond reason" (Henig, "Darwin's God," paragraph 2) is, of course, simply his *belief.*

claims about that trust. For diehard atheists that confidence may be vanishingly small but if they someday managed to get the slogan changed to, "In No God Do We Trust," although there would be many who resist the implications, the probabilities associated with that trust would merely flip-flop. The point is that either statement really covers a spectrum of beliefs. That means someone (perhaps everyone) must be wrong about the specifics. Given everything that has been said both for and against God through the years, it seems most likely that some of those perspectives are more correct than others. An inescapable consequence of the differences in such beliefs is that some folks are really promoting their ignorance. As noted in the previous chapter, however, parading one's arrogance is even worse.[4] None of this is meant to suggest that there is no religious truth to be found but I think we can safely say that doing so has not been all that easy. Later, I'll discuss why I think that might be the case but for now let's just note that our objective in religious faith, as in all types, is to maximize the accuracy of our beliefs and that doing so will require the mind of an explorer.

But where should we explore? Bumper sticker philosophy is sometimes revealing. A few years back I saw one announcing that, "God is too big to fit inside any one religion." For those who believe in God it almost sounds good, doesn't it? Most theists would probably admit to admiring a God with attributes that exceed their own. However, although we cannot presume to know the sentiments of the person displaying the sticker, it seems that this is more a statement about religion than it is about God. Such an outlook sees religion as a box that limits God and, hence, contains a sharp reproach for religion. No doubt there is some (perhaps much) truth in that perspective. But religion can also be viewed as a window through which one glimpses God and this looks like a much more constructive viewpoint. If there is a God, some windows are bound to be clearer than others. The explorer, then, is looking for just those windows that provide the best view. Faith in this or that perception of God can be buttressed or undermined based on what is or is not seen when looking through various windows. An offhand rejection of religion pulls the blinds on all such windows and leaves one with few alternatives if the primary purpose is to discover the truth about God. This is not meant to suggest that all religions want their adherents to find truth. But is it any wonder that truth-seekers who refuse to look through religious windows see scant evidence for God?

For several of my children, their first word was, "momma." I tried to take this like a man but was amused to note that they showed no particular

---

4. As the proverb has it, "Pride goes before destruction, a haughty spirit before a fall" (Prov 16:18). It may not generally be appreciated that it applies to the religious as well as the irreligious.

discrimination when saying it. In fact, for a few weeks, everything was, "momma." Decisions to ignore religious viewpoints are sometimes made by people who think them all identical but such individuals are like the child who can speak but one word. In both cases it may be all they know but children progress rapidly past that stage. Unfortunately, some people never extend their religious vocabulary. For them, all religions say the same thing. It always surprises me to hear this claim. Surely there are aspects that all religions share just as there are traits that all people have in common, but anyone who has made even a cursory study of religions knows that they differ in such fundamental ways as the attributes of deity, the place of humans in the overall scheme of things, how one relates to deity, and even whether there is a deity (or multiple versions of it).

Given such differences, is there any useful common ground from which we can proceed? I suggest that there is and that it starts with what we should not believe. In the first place, we should not feel compelled to believe everything that any particular religion tells us. There is apt to be a mixture of truth and error in any religious perspective of any consequence (including what is or is not of consequence). If there is a God, we can expect some religions to have a better view than others but, if there is not, then a religion such as Buddhism may be closer to the truth. As a corollary of not being under obligation to believe everything we are told, neither must we believe that any one religion has everything all worked out. Religious historians know that every major religion has gone through transition periods and some flux is probably inevitable for the life of any religion. This means that even those who are closest to the truth have been and will continue to be changing. I'm pretty sure this will offend some people who would prefer to believe that their particular religion is completely stable but that is neither true nor desirable. It is not true because, while it might be the case that the God of one's religion is unchanging, the collective view will not be if for no other reason than it is an amalgamation of many different individual views. It is not desirable because a static faith means there is no growth in theological insight, which seems particularly damning for anyone who professes to believe in a God with infinitely superior attributes. If we already know all there is to know about God, he must not be. If we believe we have learned all we are capable of learning, we have too low an opinion of ourselves. On the other hand, if we think we know all that God wants us to know, our opinion of ourselves is too high.

The one thing we absolutely must believe is that at any stage on our journey we could be wrong about our religious beliefs. This doesn't have to mean that we lack confidence, just that we joust with humility on a regular basis. Only then do our faulty beliefs stand a chance of being unseated. It is revealing to note that Jesus leveled one of his more scathing rebukes at religious leaders who thought they had reached the end of their religious quest but his words are germane to the practitioners of any religion (including Christianity): "If you were blind, you would not be guilty of sin; but now that you claim you can see, your guilt remains."[5] Or, if I might paraphrase, "Any claim to possess 20/20 spiritual vision is not only premature but the epitome of conceit."[6]

As we saw above, one religion will be more accurate than others on some—perhaps all—issues so there is no reason for us to believe that all religions represent barriers to truth. It may be the case that they do, but it is much more plausible to think that each contains elements of truth, instead.[7] A meaningful task, then, becomes finding the one that is most accurate. For that we should consider ourselves free to use whatever other tools are at our disposal (and to acquire those tools if they are not). In fact, the final thing I will suggest that we do not have to believe is that the other (supposed non-religious) ways of looking at the world are in competition with religion or are unrelated to it. For example, the perspective popularized by Stephen J. Gould that views science and religion as independent areas of inquiry[8] is both unnecessary and unfortunate because it assumes that neither area has anything substantive to contribute to the other. In the spirit of this section I willingly acknowledge that his view could be correct but there is little reason to accept such a proclamation as gospel with any less scrutiny than we would give to an explicitly religious claim.

The Nature of Religious Belief

Most people have a distinct impression that religious faith is somehow different from other types of faith. In fact, the vehemence with which religious beliefs are attacked or embraced suggests that they are grossly unlike

---

5. John 9:41. There is an interesting physical analog to this in a deficit called Anton's syndrome where patients who are cortically blind nevertheless deny their disability (Heilman, "Anosognosia," in Prigatano and Schacter, *Awareness of Deficit*).

6. As noted in the last chapter, this general principle is applicable to any search for truth and is in no way restricted to religion. If there is anything to a religion, however, the consequences of such pride may be more significant.

7. As C. S. Lewis put it, "If you are a Christian you do not have to believe that all the other religions are simply wrong all through ... As in arithmetic—there is only one right answer to a sum, and all other answers are wrong: but some of the wrong answers are much nearer being right than others" (*Mere Christianity*, 31). I am not prepared to say if an Islamic or Jewish analog to Lewis would say something equivalent regarding their particular belief system but I hope they would.

8. Gould, "Nonoverlapping magesteria."

anything else we might believe. In one sense that is quite wrong. As we've seen, faith is an assessment of the likelihood of something being true and that includes religious faith. Besides, the same basic cerebral machinery supports all of our beliefs—religious, scientific, relational, you name it. Claims for a genetic propensity to religious belief,[9] therefore, seem problematic. If there really is a "God gene," one might wonder why there would not also be genes for other categories of beliefs with resulting predispositions to believe (or disbelieve) that politicians are evil, scientists can always be trusted, actors are experts on the environment, and microorganisms are out to get us. I suppose we could concoct names for these hypothetical genes but there is little reason to assume their existence. As we've already noted, brains adopt beliefs of all kinds (including religious ones) because belief is what brains do. Consequently, there is no need to postulate a special mechanism for religious faith or to think that the types of things religious or irreligious people believe imply those individuals are a genetic anomaly.

Furthermore, although the actual depth of and justification for a religious belief can be significant, it is often much easier to criticize what we would like to think someone believes than it is to take the trouble to find out what they really believe. C. S. Lewis characterized this approach by noting that some

> people put up a version of Christianity suitable for a child of six and make that the object of their attack. When you try to explain the Christian doctrine as it is really held by an instructed adult, they then complain that you are making their heads turn round and that it is all too complicated and that if there really were a God they are sure He would have made "religion" simple, because simplicity is so beautiful, etc.[10]

I don't think this is the only reason people give for creating their straw men, but one doesn't have to search very far to find that stances of this sort are not uncommon. Certainly they are not restricted solely to unsophisticated views of Christianity nor are Christians immune from adopting an analogous posture. People can assault first-grade views of any religion and this is a typical behavior when dealing with one that is not our own. It is also true that, on occasion, religious people who would be offended if someone tried to undermine their religion by attacking a naïve version of it do the same thing with respect to various scientific issues. Some anti-evolutionists,

---

9. Cf. Kluger et al., "Religion."

10. Lewis, *Mere Christianity*, 36. Note that the complexity of major theories in the sciences is, likewise, lamented, but few seriously believe that such complexity undermines their significance (cf. ibid.).

for example, point to the obvious visual differences between a monkey and a human (that any six-year-old could see) and hold that up as evidence that evolution is a bogus theory. The age of the earth, fossil evidence, genetic similarities, and so forth are just too "difficult" to deal with. This problem can even afflict a person with respect to her own religion. In this case, alternative views on some theological question are met with disdain and criticized with disparaging remarks that have nothing to do with an honest attempt either to understand the issue at hand, how the other perspectives might shed light on it, or what the relative merits are of the various approaches.[11] In short, the potential disconnect between what a religious person believes and what other individuals believe about what that person believes—fostered by the sound bite mentality of our culture—often leads to the regrettable result that communication about potentially important issues totally breaks down. I hope we can avoid that here.

Although there are many directions from which to approach the nature of religious belief, my goal in this section is to consider why miracles and sacred texts, two of the primary areas in which religious belief has traditionally been assumed to germinate and grow, can function as either stumbling blocks or stepping stones to rational religious faith. What one believes about the existence and attributes of deity, involvement of deity in nature, relationship to deity and others, meaning and purpose, or an afterlife are often associated with what is believed about these.

Miracles

The question of miracles is the question of God's interaction with the world. We are thus led to look at what occurs and ask if that which we see indicates a hands-off kind of God, an occasional tinkerer, a micro-manager, or something altogether different. Furthermore, because actions and attributes are typically intertwined, conclusions here will invariably serve as a gateway to beliefs about the type of God involved. Here we are also faced with the perplexing problem of how to reconcile the good and bad things that happen with any particular conception of God (or whether we can infer anything useful at all).

Before proceeding, however, it is helpful to note that although perceived inconsistency in the attributes or behaviors of deity has been a sticking point for many people through the ages, rejecting God on that basis alone is far too simplistic. If our predecessors had done that in science they would have abandoned electromagnetic theory because wave and particle explanations

11. Cf. Smith, *Bible Made Impossible*.

were inconsistent or rejected quantum theory because randomness and determinism seemed incompatible.[12] Analogies are not evidence, but they do suggest that attempts to piece together observations under a single coherent interpretation can lead to insights that are far more productive than lamenting their apparent irreconcilability. Lack of understanding is seldom a good reason for snubbing something. Rather than throwing in the towel on the question of God's actions and characteristics (and, ultimately, his existence), persons trying to understand these issues through the lens of religious belief are faced with the far more difficult but potentially illuminating task of seeing what can be made of the evidence.

But what counts as evidence of God's interaction with nature? In many religions, miracles have been the traditional means by which deity reveals itself. In this regard there are two key questions that we must try to answer: (1) Did the reputed event occur? (2) Was it a miracle?[13] The first is addressed by the methods of historic inquiry. If the probability is low here, there is little reason to consider the second but we need to acknowledge that these are not necessarily independent. We might be inclined to rank the occurrence of an event as unlikely precisely because we have already decided that it is a miracle that we are not willing to allow. In any case, if we grant that an event occurred, deciding if it is a miracle is probably going to be very difficult because trying to get a consensus on what constitutes a miracle has thus far been next to impossible.[14] Furthermore, even if we reach a consensus, it does not remove the need for interpretation, and there's the rub. In other words, we might concur on the definition of miracle and even agree that miracles are the calling cards of deity, yet interpret what we see so as to automatically include or exclude them, thereby making the existence or attributes of God all but certain or seemingly impossible.

For example, consider two individuals who agree that a miracle is a violation of the laws of nature. Imagine that they both happen to observe a rare event and are in accord as to what occurred. One labels it a miracle and sees it as confirmation of God's existence and his involvement in nature. The other has no problem with the (mutually acceptable) definition of miracle

---

12. Cf. Bunch, "Two Stories."

13. Not surprisingly, this is similar to what takes place regarding any matter of belief. As we discovered when first trying to define faith, we are typically engaged in trying to answer two kinds of questions: "What?" and "What does it mean?"

14. Cf. Ward, *Big Questions*. Ward contrasts his own definition of miracle: "an extraordinary manifestation of spiritual power" (91) with that of Hume: "a transgression of a law of nature by a particular volition of the Deity, or by the interposition of some invisible agent" (*Human Understanding*, section X, 181) and Swinburne: "an event of an extraordinary kind, brought about by a god, and of religious significance" (*Concept of Miracle*, 1).

but argues that there was none for the simple reason that nature's laws are never violated. For him, miracles cannot and therefore do not occur and anything that passes for a miracle in the opinion of others, while it may be unusual, is not outside the realm of what nature can do on her own without external help. Because both observers saw the results of something that happened in the natural world, there is no easy way to verify that nature's laws have or have not been violated.

You might think that the second individual in our example is the person who eschews deity but it doesn't have to be. God, for that person, might be so intertwined in nature as to make the distinction between natural and supernatural, shall we say, unnatural. That may or may not be a rare perspective but later I will discuss how we might learn something important from it. In any case, a person inclined to see the activity of God throughout the natural order, rather than dismissing miracles altogether, is probably more likely to define miracles differently—as labels for rare but possible events in the normal operation of nature. Someone else, however, accepting the same characterization of miracles (i.e., rare but natural events) will conclude that God is not needed—nature alone is sufficient. Others might see the order in nature itself as something of a miracle.

As we've already seen, how we interpret evidence or even what we are willing to call evidence is subject to what we already believe. So, whether we adopt a forceful or watered-down definition of miracle, it may make little difference if we have previously determined whether they can or cannot occur or what they signify if they do.

For some people, then, miracles serve as evidence on which their religious faith is built but branding an event as a miracle and attributing it to deity are also acts of faith. This can create a self-supporting cycle in which designating something as a miracle buttresses one's faith which in turn makes that person more likely to see something else as a miracle, and so forth. This is not unusual in matters of belief—for example, a person who attributes an action by another as a token of affection might be led to believe she is loved by the latter, making her more apt to interpret subsequent actions as signs of endearment. That may be the case or the actions may simply be due to the courteous nature of the one performing them or even coincidental. Potential evidence always requires careful handling but proper interpretation of reputed miracles seems especially important because what is at stake may be no less than a proper belief about deity.

It is interesting to consider the range of views on miracles by people in modern societies. I've identified some alternatives above but those were just broad categories. In the case of individuals who believe that God intervenes in nature in special ways one can find people who accept miracles as possible and believe they occurred in the past, but who don't believe they occur very often, if at all, any more. Others stake their lives on them, refusing medical help in times of crisis, believing that if God wants to spare them, he won't need a physician's help to do so.[15] Such attitudes not only reveal the conflicted way in which we are prone to consider potential evidence, they expose something important about our real beliefs with regard to how God might intervene in nature.

Sacred Texts

People primarily know about the miracles on which a religious faith is based through the sacred scriptures of the respective religions. Sometimes reports of those miracles apparently come from people quite close to the source but on other occasions the event is recorded and possibly interpreted long afterwards. Believing that the author got it right may be based on an accompanying belief that the information was received via divine revelation. I have already noted that revelation can be interpreted in various ways but this suggests another door through which deity might enter into nature and it will be viewed by some as a special form of miracle.

As with perspectives on miracles in general, our views on revelation reflect what we think God is like and how he operates. The same circular form of support for our beliefs is operative here (i.e., what we think God is like and how he operates will also determine our views on revelation) but we might wonder, "With which does one start?" That is, do we begin believing in revelation and then use a purported example to shape our belief about God or does an existing belief about God shape our interpretation of a supposed revelation? Alternatively, do we begin by disbelieving in revelation and consider an apparently phony example as confirmation that God doesn't exist or does an existing lack of belief in God cause us to discount any and all claims for revelation? As you can see, this chicken and egg dilemma is no respecter of theistic or atheistic position. In either case there must have been something that originally disposed us to one view or the other and, therefore, in the process of crafting a rational faith (as it pertains to belief in the existence, attributes, and behaviors of deity), we should be interested in what that might be. Identifying such a starting point, however, can be quite difficult. Fortunately, it may be less important to pinpoint the

---

15. Charlatans sometimes prey on the holders of such beliefs, discrediting for many the possibility of genuine miracles. However, this is but one form of swindling which is propagated by people who would undermine a legitimate faith or reinforce a false one just to serve their own selfish interests.

THE NATURE OF RELIGIOUS BELIEF   181

actual origin of our disposition toward a certain mode of belief than it is merely to recognize that there was one and that it may or may not be sound.

A real problem here is that although understanding God in some religion is supposed to be based on its sacred writings, determining what constitutes those writings and what significance to afford them is in large part dependent upon beliefs about how God relates to humans. We've already seen how belief cycles such as this are common but it is of special importance in this case because the authority of the sacred writings themselves (or some portion thereof) is at stake. Did God really interact with humans in such a way as to make himself known? If so, a record of that interaction should be terribly important. On the other hand, if someone is mistaken about the source or content of an event or insight, it would be important not to allow it to achieve the status of revelation.

A good illustration of the conundrum involving what to count as sacred can be seen in one of Paul's letters to the Christian congregation at the Greek city of Corinth. Consider his instructions regarding marriage:

> To the married I give this command (*not I, but the Lord*): A wife must not separate from her husband. But if she does, she must remain unmarried or else be reconciled to her husband. And a husband must not divorce his wife. To the rest I say this (*I, not the Lord*): If any brother has a wife who is not a believer and she is willing to live with him, he must not divorce her. And if a woman has a husband who is not a believer and he is willing to live with her, she must not divorce him.[16]

At the beginning of this passage, Paul believes he is merely passing on the instruction of Jesus.[17] In the subsequent instruction, however, he explicitly acknowledges that he is not repeating anything Jesus said. As a result, everyone who reads this, beginning with his original audience, must decide if Paul is speaking for God or not.[18] Some would say the very fact that his words made it into the collection of Christian sacred writings means he must be relating God's intentions, but others aren't so sure. His advice may be good, but did God give it to Paul without Paul being able to say confidently that he did? In the verses following, Paul goes on to suggest that marriage is probably best avoided for "the time is short" and "this world

---

16. 1 Cor 7:10–13, italics mine.
17. Mark 10:2–12.
18. Of course they must also decide if Paul properly interpreted Jesus' position, if that position was correctly recorded, and if there is any particular reason to think Jesus' insight deserves special consideration. Belief is operative in each of these considerations.

in its present form is passing away."[19] Folks can interpret these sentiments in any way they wish to make them fit whatever they want to believe Paul meant, but the most obvious explanation is probably that Paul was mistaken about the imminent return of Christ. Should he, perhaps, have qualified these statements, too, with, "I, not the Lord"? The variety of interpretations given to his instructions in this chapter are apparent in the contrasts between the monastic lives adopted by those who take his advice at face value and the tendency of most others (including ministers who otherwise claim to take the Bible literally) to find an interpretation that makes it possible for them to justify their own marriages. It is interesting to note that Paul concludes the section on marriage by noting that he thinks he has God's Spirit,[20] which can be read as a sarcastic way of saying of course he does or as a humble acknowledgement that he might not have gotten the message as clearly as he would like to think. Confidence is frequently commendable but over-confidence is no virtue and, as we saw previously, admitting that our beliefs can be in error is fundamental to growing a rational faith. Perhaps this is even true for apostles.

The kinds of observations I have just made sometimes rankle those who have unquestioningly accepted the authority of the sacred writings of their particular religion, but that doesn't mean that authority is wrong any more than their willing acceptance means it is right. However, it needs to be perfectly clear that when the primary source which describes the interaction between deity and humans is presumed to be part of that very interaction, we should be prepared to look deeper for evidence that such a claim really has merit and is not merely self-serving.

The need for critical evaluation doesn't stop there, however. Sermons, lessons, hymns, books, and related forms of religious instruction are usually intended to clarify the sacred writings, but in one sense they compound the problem because they reflect the beliefs of preachers, teachers, or authors about what is important enough to deserve attention and how it should be interpreted. Recipients of such messages must then determine not only if God spoke through the particular writings being examined but if he also interacted with the person conveying the message firsthand (and, if he did, whether that person got it right). Most theistic religions postulate that the very process of trying to determine the nature of God's interaction with humans via such examination constitutes a part of that interaction, suggesting that the whole enterprise is a very dynamic one.

---

19. 1 Cor 7:29, 31.
20. 1 Cor 7:40.

Beyond belief?

According to the Colorado Fourteeners Initiative,[21] there are fifty-four peaks in that state that rise to a height of, well, fourteen thousand feet or more. I climbed my first one in 1983 and have been slowly knocking off the others on repeated trips to the Rockies. Over the past twenty-eight years I have ascended fifty-one of them—not too bad considering the closest one is over 1,000 miles from my home, but I'm not bragging. You see, one Ted Keizer has climbed all fifty-four in less than eleven days.[22] Several years ago, a friend and I drove to Wyoming to climb the Grand Teton. On the first day we backpacked to a high camp, did the technical pitch on day two, and returned to the car on day three. In 1983 a fellow named Bryce Thatcher did the climb (which tops out at 13,770 feet after a 7000 foot elevation gain over about fifteen miles roundtrip) in a tad over three hours.[23]

I like to think that a more judicious use of vacation time or a residence in Colorado would have allowed me to climb all fifty-four of the fourteeners in a much shorter period of time than it has taken thus far. But eleven days? No way! For almost anyone who has ever attempted to climb one of those mountains, particularly if it is one of the more difficult peaks, Keizer's feat is all but unbelievable. I've also entertained the prospect of trying to summit the Grand in one long day but am no more deluded about duplicating Thatcher's accomplishment than I am about growing wings (which might actually make it feasible).

I have chosen these humbling examples as a reminder that, in practically any area of life we might care to ponder, there are things happening that stagger the imagination. From inconceivable numbers of consecutive pushups and outlandishly long but survivable falls to flashes of scientific, literary, or musical genius that leave us gaping in amazement, we are confronted with performances that surpass the average person's credentials to such an extent as to almost seem divine. Of course no one is really about to attribute deity to any of these supermen or superwomen if for no other reason than that their outstanding performances are invariably accompanied by a host of obvious frailties. But there are several other reasons that we stop short of that designation.

In the first place, many of these accomplishments are not achieved in temporal isolation but are the result of a gradual increase in performance capacity. For example, the current world record for the mile is incredible

21. http://www.14ers.org/.

22. Lanza, "Speed Freak," 82. Besides those fifty-four, there are numerous subpeaks over 14,000 feet.

23. Ortenburger and Jackson, *Climber's Guide*.

compared with the capability of an average human but when it is seen as a steady lowering of times over many years by numerous dedicated athletes (and even for individual athletes) it appears feasible for mere mortals (or at least for demigods!). We can even almost imagine ourselves running that fast under the right circumstances, although most of us know that we never really could. Furthermore, running is an extension of something almost all of us do; so despite the incredible abilities we see in Olympians and the like, they are at best incredible with an asterisk. We would consider it far harder to believe that someone could fly unaided than that they could run extraordinarily fast simply because we know how to run and we know people who know how to run but flying is not something humans do. Naturally, we know that a small minority of people can run with tremendous speed because we have seen them do it or have read reports of their exploits. In other words, we consider their accomplishments to be well attested—which is just another way of saying that we are satisfied with the evidence.

For all these reasons, we are in little danger of deifying superstars (assuming we disregard the large financial offerings we make to their coffers, the pilgrimages to view their performances, etc.) but it is worth reiterating that many of their exploits are so far outside the range of our personal experience (i.e., in terms of what we are capable of doing) as to appear almost miraculous. In other words, for most of us to do those things, it would take a miracle. You might wonder, then, how you would respond to reports of a person who didn't just excel at running or climbing or composing or scientific thinking but was superior in all of those areas—or if you suddenly saw someone fly. Would you disbelieve your eyes, think that person a god (or an alien), or wonder about performance enhancing drugs?

We all have ideas about things we deem possible and impossible plus some about which we haven't yet decided and it is an instructive exercise to consider the process by which things move from one category to the other. As we've seen, however, because faith exists on a continuum, it is not quite as black and white as that. In an earlier chapter I asked you to perform a simple exercise in order to explore the graded nature of faith and I am hoping you will repeat that task here but with a different set of statements (and a different goal in mind). Your assignment is to examine each of the claims below, ranking it on a scale from 0.0 to 1.0 according to the extent to which you accept it as true (0.0 = absolutely not true and 1.0 = absolutely true):

1. It is possible for a woman to become pregnant without sexual intercourse.

2. Dead people can be restored to life.

3. Blind people do not have to remain sightless.
4. Someone could start with a handful of bread and fish and feed thousands.
5. The universe was created from nothing.
6. Under the right conditions, life can arise spontaneously from inanimate elements.

If you take this exercise seriously, I imagine that there may be a bit of hesitation as you try to decide what is really meant by each statement. You may be asking yourself, for example, "Does he want me to indicate whether I believe that, some two thousand years ago, a Jewish teenager became pregnant without sexual intercourse, or is he asking about in vitro fertilization?" You could easily assume that I am asking about religious miracles but perhaps you're also trying to recall the current medical definition of death or wondering if cardiopulmonary resuscitation or cataract surgery count. I expect you noticed that all of the items in the list have religious connotations but it should be equally apparent that they need not. In any case, I hope that you are beginning to wonder just what it really means for something to have religious implications and why we might decide that one event does but another does not.

There is sometimes a tendency to associate religion with the ethereal but all of the items in the list above pertain to the physical world. Nevertheless, all of them are currently outside the experience of some humans and some are outside the range of experience of all of us. There was a time, of course, when virtually all of these would have been considered miraculous but, as we saw when considering "God of the gaps," an ability to suggest a mechanism has often had the effect of removing God from the picture. This seems to have been the basis for Laplace's response to Napoleon when asked about God's role in his theory of planetary motion, "I had no need of that hypothesis."[24] In reality, however, Laplace implicitly acknowledged to Napoleon (and us) the limits of his own inquiry—it is only possible to eliminate God if one has already determined how God works.

Ironically, religious people have been as guilty as anyone of propagating the perception that God's activity must have a mysterious air about it and many remain unduly bothered by attempts at explanation. But religion is not (at least it doesn't have to be) the neat little box too many (including religious folks) have presented and religious perspectives on presumed cut-and-dried issues can show great depth. William Barclay, for instance, suggests that the incident in which Jesus took a lad's meal and fed thousands

24. Quoted in Bell, *Men of Mathematics*, 181.

was accomplished because he was able to leverage one example of sharing into encouragement for others to do so.[25] I suspect this interpretation will insult persons who prefer to think that a "natural" explanation undermines the significance of the event and am equally sure that it will be received gladly by those who knew all along that nothing "supernatural" really occurred. But both reactions involve presumptions no less restrictive than Laplace's. Which, after all, is the greater miracle: mysteriously multiplying food or changing human hearts (as Barclay proposes)?[26]

It is no good complaining that one group of possibilities is "natural" and the other "supernatural" unless you are willing to say that you know how to draw the line. For the ancient Israelites there was no clear distinction since God was involved in everything and, therefore the supernatural was not necessarily unnatural.[27] Barclay thinks this has some merit[28] (and so do I) but too loose a merging of the natural and supernatural is fraught with several problems, not the least of which is discriminating religious beliefs from any others. Is God really involved in everything or in nothing? What should count as a religious belief? Perhaps this is a predicament only for linguistic nitpickers but I think it runs much deeper. Historically the dilemma with attributing everything to God is that it leaves no room for freedom as far as we can see. The real problem is that we cannot see very far, although poor vision has not prevented many from describing what they wanted to see.

Despite the alternative ways to think about God's presumed relationship to nature (and perhaps precisely because we are thinking about that), we cannot evade the nagging question: Are some things just unbelievable no matter what? More to the point, are all of the attempts to characterize God's involvement in nature misguided simply because God himself is beyond belief? If religious people believe things that are, in principle, impossible, then religious faith can never be rational. But if I believe that some things are not believable, is that not equivalent to saying that I believe it to be impossible for *me* to believe them. Isn't that, then, a belief in the impossible? Maybe this is an issue that transcends religious belief. No doubt some things really

---

25. Barclay, *Gospel of John*. See John 6.

26. Barclay even makes a good case that the story in Matt 17 (regarding finding a coin in a fish's mouth)—which many have assumed was a miracle—was likely the result of an expression used by Jesus to make a point (*Mind of Jesus*). Barclay's point is that all miracles are not created equal. Whether this is a slippery slope or a way to climb out of a "crude and humorless and uncomprehending literalness" (ibid., 84) is for the explorer to determine (but not pre-determine). In any event, Barclay is not suggesting such an interpretation for all supposed miracles.

27. Barclay, *Mind of Jesus*.

28. Ibid.

THE NATURE OF RELIGIOUS BELIEF 187

are beyond belief (or should be), but how does one decide that something is unbelievable? We've already seen that we must be careful about letting our own experience define the limits of believability but a lack of experience is no excuse for indiscriminate belief, either. There might be a tendency to let science define the boundaries of believability, but if science has shown us anything it is that those boundaries are flexible.

On the one hand, then, people confronting certain religious claims— particularly those involving the interface of super-nature and nature—are constrained by the bonds of their own limited vision while on the other they need an unfettered vision in order to see clearly what to make of those claims. Ostensibly, this is an unsolvable dilemma because those bonds are unbreakable (and we can expect that our vision will always be limited). However, that doesn't mean they can't be stretched. We may not see everything, yet there is no reason to think our vision cannot be improved. How much insight is ultimately possible, no one knows, but the minimum requirement is to be able to see the constraint in the first place.[29]

Unfortunately, some of the shackles are of our own making. For example, it is impossible to stretch the limits of comprehension by simply defining away perceived problems but people insist on doing so nonetheless. Thus, "natural" and "supernatural" are sometimes defined so as to preclude certain possibilities. If we define "supernatural" as that which is beyond our ability to explain, then of course it will appear to dwindle away as our understanding grows. But who determined that the supernatural must always operate outside our explanatory powers?[30] Who decided that, if we can explain something, it is it then no longer supernatural? Allowing someone to define things anyway he pleases makes it possible for him to prove anything he pleases. Ideally, we will understand "natural" and "supernatural" in a way that neither unduly restricts deity nor is so general as to make virtually anything possible. Although it might be true that anything is possible with God, such a claim in isolation offers little in the way of explanatory content. Whatever the definitions, they function as the assumptions by which the remainder of any theological edifice is understood and so, as with any assumptions, open or shut the door on further insight.

---

29. It should be apparent that lack of understanding on our part could only bound a puny God (that is, a God whose attributes have already been determined to be limited).

30. Some people even maintain that, "theism provides by far the simplest explanation of all phenomena" (Swinburne, *Is There a God?*, 41).

What We Don't Yet Know

All of this is irrelevant, we might surmise, for anyone who has already decided that God's ways are unknowable or who has predetermined the boundaries of his or her own understanding. After all, what can we really expect to know about God's intervention in nature? Isn't the very idea that mere mortals can understand something of the Almighty's actions a bit presumptuous? Certain religions may even seem to foster this idea. Consider, for instance, the Psalmist's stance:

> My heart is not proud, Lord, my eyes are not haughty; I do not concern myself with great matters or things too wonderful for me.[31]

Pardon my impudence, but if one isn't concerning himself with "great matters," what is of concern? And how, pray tell, does he know what is (or is not) too wonderful for him? I applaud his humility but question his assumption because it stakes him to a limited understanding from which he is unlikely to move. This is not just any old assumption but one which, if believed, inhibits further development.

The Psalmist's position sounds a bit like Thomas Jefferson's comment to John Adams in 1820 that, "When I meet with a proposition beyond finite comprehension, I abandon it as I do a weight which human strength cannot lift, and I think ignorance, in these cases, is truly the softest pillow on which I can lay my head."[32] But, just because something gives the appearance of being beyond my or your or even Jefferson's comprehension, is it therefore beyond that of all others? Could it be that something is only beyond our comprehension because an effort to understand was never begun or was prematurely aborted? It is no denial of the need for humility to note that the conclusion that something is beyond one's comprehension is a matter of faith.

In any case, I suspect it is a mindset typified by the Psalmist to which George Luger referred when he said that, "The notion that human efforts to gain knowledge constitute a transgression against the laws of God or nature is deeply ingrained in Western thought."[33] It is debatable just how ingrained that notion is, but one doesn't have to look far to see evidence that it exists. Unlike the Psalmist, Luger is not advocating resignation, but both of them

---

31. Ps 131:1.

32. Cappon, *Adams-Jefferson Letters*, 562. Thanks to historian Brian Steele for bringing Jefferson's position to my attention and also for noting that, "Jefferson was definitely not one to otherwise foreclose possibilities" (personal correspondence).

33. Luger, *Artificial Intelligence*, 4.

have apparently overlooked the plea (also a part of Western thought) to seek knowledge, found just a few pages after the passage above:

> Indeed, if you call out for insight and cry aloud for understanding, and if you look for it as for silver and search for it as for hidden treasure, then you will understand the fear of the Lord and find the knowledge of God. For the Lord gives wisdom; from his mouth come knowledge and understanding.[34]

Perhaps we can sort this out via a simple exercise. To begin, imagine a horizontal scale representing the quantity of all possible knowledge. The left end of the scale is fixed and corresponds to knowing nothing at all. Because we expect the sum total of all knowledge to be immense but don't actually know its extent, the right end disappears somewhere in the distance. Your mission is to position the following items on the scale:

- How much you currently know
- How much it is possible for you to know
- How much God knows

I suspect that the results for many people will have the general character of Figure 1 but if yours is different, so be it.

Figure 1: *Knowledge by the numbers. Thinking about what we still can learn . . .*

For the Psalmist (quoted above), the left hash marks are dreadfully close together, if not coincidental. However, although it may seem hopeless to think we can know what God knows, if there is any gap at all between those two marks it is not hopeless to believe that we can know more than we currently do. Furthermore, while the presumably large distance between the ultimate limits of our knowledge and God's may highlight how much we can

34. Prov 2:3–6.

never know, the truly significant point is that if God's knowledge is indeed vast, even a small fraction of that knowledge could likewise be great.[35]

I think that one reason many individuals (religious or not) have adopted the mechanic's approach to their religious beliefs is that they have failed to realize just how big the gap really is between what they think they currently know about God and what they could conceivably know. Consequently, they have seen no need to explore the contents of that gap, even though it could contain just the insights they need for the questions they have prematurely assumed were unanswerable. One might think that this would be adequate incentive to investigate but, as we noted earlier, exploration is not without its dangers. If, for the moment, we ignore the rightmost label from Figure 1, considering only the difference between what we know in general and what we could potentially know, we must acknowledge that, whether we currently believe in God or not, we might discover evidence to the contrary. On the other hand, if the gap between the rightmost hash marks is very big, then, no matter what we believe about God, he is not entirely the God we think he is. The mechanic will find this difficult to acknowledge but the explorer is mindful that what he does know is probably in need of revision.

Resistance to the idea that we can learn about God through our own rational search is somewhat curious given that (1) this is the only means at our disposal (i.e., within our power) and (2) some scriptures actually seem not only to enjoin just such a search but to promise results as well:

> Ask and it will be given to you; seek and you will find; knock and the door will be opened to you. For everyone who asks receives; the one who seeks finds; and to the one who knocks, the door will be opened.[36]

The biblical book of Proverbs is full of injunctions to obtain wisdom and knowledge. Asking, seeking, and knocking look like significantly more productive approaches than lying on the couch awaiting an unearned revelation. Why would God be interested in giving deeper insights to someone who lacked the motivation to seek them? I think the real sticking point—the reason many people have and perhaps continue to avoid rational

---

35. The imagined large distance between what we can know and what God knows is not an argument for the existence of God (the exercise assumes his existence) although, as we've seen, it has sometimes been used for that purpose. As we've noted, although belief in the impossible is not commendable, deciding what is or is not impossible may not be straightforward. Believing that there are some things we just can't know seems to be another form of belief in the impossible but in this case the restriction is on us, not God.

36. Matt 7:7–8. Also recall Prov 2:3–6.

exploration—is a concern that it somehow obviates the need for God's participation or that it smacks of hubris. Both are possible, certainly, but neither needs to be a foregone conclusion. It would certainly be presumptuous to assume that by our own efforts we can know everything God does—at least any God worth knowing—but we need not go there. There is thus no reason to assume that we are overstepping our bounds by contending that God reveals himself to those who seek him.[37]

* * *

Our primary interest in this chapter has been to explore how religious faith develops and operates, to consider its rational content, and to begin to ponder what it means to construct a set of coherent beliefs. Unfortunately, we are often prone to accept or reject things with the smallest excuse, sometimes (but not always) driven by what we do or do not want to be true; sometimes (but not always) failing to consider evidence and reason. Attempts to develop a rational religious faith can lead to apologetics or apostasy but the promised rewards (in each religion) are often too important to excuse carelessness when investigating them and the whole matter calls for special attention and care. In fact, the primary reward for the explorer is truth—not what someone imagines to be so but what is so; not what someone thinks the new world is like but what it is like. For Lewis and Clark it was not what they believed lay between the Mississippi and the Pacific but what was really there. It is not what a physician might speculate as the cause of or cure for cancer but what it really is. And it is not what God might be like or what we wish he were like but what he really is like.[38] If this comes across as a defense of deity, forgive me—God probably needs my help a good deal less than the Incredible Hulk needs SpongeBob—but I *am* defending the idea that preconceived notions, failure to consider evidence, and satisfaction with the status quo put any real knowledge of God in jeopardy for people who call themselves religious just as much as it does for those who do not.

37. Jer 29:13: "You will seek me and find me when you seek me with all your heart." I'm not suggesting for a minute that a person has to understand how God does something in order to believe that he does it any more than a person has to understand quantum tunneling to appreciate and use a computer. But is the objective of a religion just to use God?

38. It may be inconvenient but it is important to distinguish what we want from what we should want (i.e., truth).

─── CHAPTER 10 ───

What is Religion, Anyway?

Neither in its impetus nor its achievements can science go to its limits without becoming tinged with mysticism and charged with faith.[1]

—PIERRE TEILHARD DE CHARDIN

Gravity Dreams

I LIVE AT THE top of a hill, a fact that immediately places the explanatory value of physics in jeopardy—in my case, what goes down must come up. But it is not the law of gravity that is in danger. What is threatened by this observation is the presumption that science has its own forms of omnipotence and even omniscience (in the form of predictability). That is, it is the idea that science always has the last word—that there is a law for everything—which is called into question. Although science can predict what will happen if I put the car in neutral at the top of the hill, it can't say with certainty whether I will do so. It can say almost precisely how many calories I must expend to climb the hill, but is mute on the issue of whether I will ever do it again, perhaps opting, instead, to settle for a diet of pizza deliveries for the rest of my cholesterol-filled days. Clearly, something more powerful than gravity is at work.

Lest you think this an anomaly, I would like to point out that, with five kids on the loose, our house is in a state of perpetual motion. So much for the first law of thermodynamics... I will confess that the second law seems

---

1. Teilhard de Chardin, *Phenomenon of Man*, 284.

secure—the resulting disorder readily confirms the continued increase in entropy—but despite the fact that science has taken us a long way, it has thus far left us short of the destination. Whether it can ever take us all the way is a matter of faith (and some wishful thinking) and many still dream of the day when science can explain everything. The resulting devotion thus makes science look more and more like some kind of god, although you would think that its current limitations would make people wary about going that far. Those who deny the existence of any god would certainly not want to do so, but that doesn't mean they can't hold science in a type of religious awe—after all, we've already noted that religious belief does not have to involve deity. But if that is so, what makes something a religious belief?

Some years ago a person I know quite well, apparently experiencing a certain dissatisfaction with life brought on by a lack of fulfillment or, perhaps, a sense that something was missing, embarked on a quest that ended up consuming him for nearly eight years. During that time his level of devotion, concentration on the sacred writings, dedicated study and meditation, attention to ritualistic detail, sacrifice of time and resources, and overall commitment were exemplary. All of this eventually culminated in an initiation ceremony at which he was finally bequeathed the right to share the light he had received and to make new disciples. Moreover, the effects were lasting—to this day, he has not abandoned his calling.

It is hard not to think of this as a religious pilgrimage but what I have described is my experience in graduate school where the focus was not theology but the sciences. My case, however, is not unique. Practically everyone can relate similar stories where their devotion to education, occupation, hobby, or family was (or is being) played out with religious fervor, often accompanied by a sense of achievement and purpose. None of this need replace God but the dedication looks terribly religious and, when loyalties are split, one may appear almost polytheistic.

I am aware that I could be accused of trying to redefine religion but that isn't my goal and I'll take the risk. Most of us have a tendency to classify people as religious or nonreligious and, up to this point, I have done so as well but now it is time to acknowledge that things aren't that simple. The behaviors I have described above are one indication of this but so are the beliefs that drive them. If deciding that a particular perspective is germane to the source of meaning and purpose or offers insight into how one should relate to others, then what would keep us from saying that any belief about it is not religious? English professor Brian Boyd, for instance, suggests that, "Intelligence and creativity are purposes that have emerged over the course of life on earth."[2] Why is that not a religious belief? If you said, "Because it omits God," you have ignored "religions" such as Buddhism and it is you who are changing the definition. Boyd acknowledges that, "Religion partakes of elements of both art and science,"[3] but I am suggesting that the sciences (and art, too, among other things) exhibit elements of religion. Perhaps it is this thought that is needed to pull our religious and non-religious heads out of the clouds.

Is Science a Religion?

The worship, prayer, ritual, and sacred literature we associate with religion all have their parallels in other domains but I want to focus here on science precisely because it is the primary area in which many have attempted to distance themselves from traditional (primarily theistic) religion. As before, I am neither criticizing science nor exalting religion—my intention is to elaborate the claim that something closely akin (if not equal) to religious faith is really quite ubiquitous. If we find evidence of that where it is least expected, perhaps it will be easier to see it elsewhere as well.[4]

When we think of worship, it is typical to envision people singing, praying, meditating, listening to homilies, and the like, but these are generally just the liturgical manifestations of an underlying attitude. Ultimately, worship is adoration, praise, and exaltation of some entity or ideology toward which such activities might be directed, but other overt responses are possible. Worship may also occur privately but, either way, it is not restricted to a single domain of our lives. The glorification of science in modern culture, for example, is hard to miss. I gave some hint of this above but I imagine you have noticed that people are regularly singing its praises and that no small number sees it as our savior.

Biologist E. O. Wilson's perspective is a case in point: "If the sacred narrative cannot be in the form of a religious cosmology, it will be taken from the material history of the universe and the human species."[5] Here, his dismissal of traditional religion is replaced in the same breath with venera-

2. Boyd, "Purpose-Driven Life," 28.

3. Ibid., 33.

4. You will notice in what follows that I do not elaborate on science and religion parallels pertaining to ritual nor to what I termed in the description of my graduate studies "sacred writings" (i.e., books and journal articles that carry the essential message of one's scientific discipline) but I expect these to be fairly obvious. For example, anyone who has taken a laboratory-based science course understands at once something of the ritual involved in science (cf. Latour and Woolgar, *Laboratory Life*).

5. Wilson, *Consilience*, 289.

tion of a sociobiological credo of which he cannot say enough. That's okay because, as with most evangelical efforts, others are there to carry the flag, sometimes preaching the glorification of science as realized in technological advance. Philosopher Daniel Dennett, for instance, calls science "a technology of truth,"[6] and transhumanist proponent Simon Young bluntly declares, "Ultimately only technology can produce a truly radical transformation of self."[7] Surely none of this is new. Over two thousand years ago the Roman poet and natural philosopher Lucretius seemed quite pleased to think that religion was giving way to a more mature view of nature.[8]

Well, perhaps you've noticed how fond we humans are of worshipping ultimates. This adulation for science is echoed in the reverent belief of physicist Lee Smolin who remarks, "All that is possible of utopia is what we make with our own hands. Pray let it be enough."[9] Although it is not clear to whom Smolin (who has little use for deity) is praying, it should be apparent that prayer, too, has its analogs in science. I imagine you might want to object that Smolin is merely expressing a particular hope in a poetic way and I agree. But what is prayer if not an expression of one's hopes, desires, gratitude, and admiration directed at that from which one expects or has received fulfillment? The scientist bends his or her metaphorical knee to nature, theory, or the process itself, praying for revelation that will make sense of their world, better their personal lot in life, and immortalize them in the only way some deem possible. That these sentiments are not uttered aloud make them no less operative—they are the prayers of every scientist's heart and they are based on faith.

Several weeks prior to writing this section I had hiked with part of my family to the brink of a rather large depression located in Canyonlands National Park in Utah. This geological feature—ironically called Upheaval Dome—is of interest not only due to the tortured terrain it encompasses but particularly because scientists cannot agree on what caused it. Whether a meteor, collapse of a former subterranean salt deposit, or something else (perhaps a clandestine excavation operation by the Park Service?), no one can say for sure, but that doesn't prevent various scientists from believing in one theory or the other. Proponents of each view will point to selected evidence to support their faith but there is no known way to resolve the issue and whatever historic event caused it cannot be repeated. This dilemma, of course, is not limited to geology (although there are plenty of

6. Dennett, "Why Getting It Right Matters," in Kurtz, *Science and Religion*, 156.
7. Young, *Designer Evolution*, 371.
8. Carus (Lucretius), *Nature of Things*.
9. Smolin, *Life of the Cosmos*, 300.

other examples in that field[10]). Cosmology and evolutionary biology quickly come to mind but any scientific theory that deals with the very ancient or that postulates phenomena beyond our current ability to measure demands a faith that looks suspiciously like that which many people think is only typical of traditional religion.[11]

But that isn't the only kind of faith the scientist exhibits. The very exaltation of science we discussed earlier is a display of faith in its powers. As the psychologist Karl Lashley once noted, "Our common meeting ground is the faith to which we all subscribe, I believe, that the phenomena of behavior and of mind are ultimately describable in the concepts of the mathematical and physical sciences."[12] This is what McNamara would call the "Western faith in scientific rationalism"[13] and rests on what Keith Ward describes as the faith needed by scientists to believe that "every event has a cause" and that "the laws of physics will apply everywhere in the physical universe."[14] In perhaps its most extreme form one finds what Victor Reppert has termed "scientific fideism."[15]

That kind of faith is the starting point but it must be followed by a more directed kind which leads the scientist to believe that he can operate productively within some specific paradigm,[16] drives him to explore one hypothesis instead of another, encourages him to accept the word of fellow scientists, persuades him to rely upon extrapolation and inference, and makes

---

10. Contemporary geological theory provides a classic case of the significant role played by inference in the interpretation of ancient formative events and their causes (cf. Meldahl, *Rough-Hewn Land*).

11. This raises some significant questions about the claim by Eugenie Scott (former Executive Director of the National Center for Science Education) that "There is only one science, but there are many religious views" ("Science and Religion Movement" in Kurtz, *Science and Religion*, 114). Perhaps she meant that, in principle, there is only one scientific method (although the exact form that should take or has taken is debatable) but there have been and continue to be many scientific views (e.g., the previously referenced Upheaval Dome debate).

12. Lashley, "Problem of Serial Order," in Jeffress, *Cerebral Mechanisms*, 112. When he said "we all," Lashley didn't necessarily mean you—he was speaking to a group of (presumably) like-minded peers. Note, however, the double level of belief in his statement.

13. McNamara, *Evolution, Culture, and Consciousness*, 98.

14. Ward, *Divine Action*, 5. John Haught suggests that, "naturalism itself turns out to be a construct erected on an avowedly *faith-filled* decision that only a method of inquiry that leaves out any reference to purpose is intellectually acceptable" ("Purpose in Nature" in Morris, *Deep Structure of Biology*, 223, italics in original).

15. Reppert, *C. S. Lewis's Dangerous Idea*, 124–28.

16. Cf. Kuhn, *Structure*.

him think he can safely ignore certain observations.[17] Small wonder, then, that Planck thought the words, "Ye must have faith," adorned the "gates of the temple of science."[18]

But surely scientific faith is ultimately different from traditional religious faith, is it not? Francis Crick admitted the role of faith in science but argued that scientific faith is "provisional."[19] His assumption is clear: religious beliefs are not. But history (if not one's personal experience) makes it plain that traditional religious beliefs are provisional as well. If they were not, there would have been no emergence of monotheism from polytheistic roots, no chance for Christianity or Islam to spring from a Jewish heritage, no Protestant Reformation, and so forth. Furthermore, religious beliefs are not the only ones that are hard to let go—recall the frequent resistance of scientists when confronted with a new paradigm.[20] There may be an inclination to think that scientists are at least in principle more open to changing their beliefs in the face of supporting or disconfirming evidence but, even if true, that is not a condemnation of religious faith per se but of how some people embody it. Who says we cannot modify our religious perspectives (one way or the other) when faced with compelling reasons to do so?[21] Unfortunately, there are persons (traditionally religious, scientists, both, or neither) who start with the premise that there could in principle be no convincing evidence. How sad . . .

Part of the evidence conundrum—one that has become a common complaint about religious beliefs from those on the periphery and many insiders, to boot—centers on the way in which the devout understand the narratives in their respective traditions. Joshua, Job, Jonah, Jesus, Jehovah: what parts of their stories should one construe as literal and what is merely figurative and

17. Kuhn's insight here is telling: "Reports of effective research repeatedly imply that all but the most striking and central discrepancies could be taken care of by current theory if only there were time to take them on. The men who make these reports find most discrepancies trivial or uninteresting, an evaluation that they can ordinarily base only upon their faith in current theory. Without that faith their work would be wasteful of time and talent" ("Essential Tension").

18. Planck, *Where is Science Going?*, 214. John Haught suggests that, "faith in the importance of truth-seeking is what initiates and energizes the actual practice of scientific method" ("Science, God, and Cosmic Purpose," in Harrison, *Cambridge Companion to Science and Religion*, 264).

19. Crick, *Astonishing Hypothesis*, 257.

20. Kuhn, *Structure*; cf. Barbour (*Religion and Science*) on scientific and religious paradigms.

21. There seems to be a tendency on the part of some people (both religious and not) to believe that any possibility for change in religion somehow invalidates the religion in its entirety but why must anyone assume that? If we harbored the same type of belief about science there would be no such thing.

instructional? It seems unlikely that all are equally evidential, and people who accept or deny them do so, as we've seen, with different probabilities for each. I happen to think this is a critically important issue that requires tremendous discernment—Barfield considers an overblown tendency toward literalness a form of idolatry[22]—but it is not a problem unique to traditional religion. Should we take electron orbits as literal, for instance?[23]

Even the circumstantial evidence that guides interpretation of some religious issues is not without its parallels in science. Should the requirement for believing the expansive claims of evolution be direct observation at a species-changing scale? Is the Big Bang only worthy of our consideration if we can find someone who saw it happen (and would we believe him)? Must we travel beside a receding galaxy in order to rationally believe that the universe really is expanding? Both the gospel and first epistle of John affirm that, "No one has ever seen God"[24] but, by definition, no one has seen a black hole either. Many objects of devotion in science and religion are of this nature, where the existence of something is inferred on the basis of its supposed effects, possible theoretical considerations, and the like. The fact that we can never "really" see it has never been much of a deterrent to belief. Perhaps that is because some criteria for sufficient evidence have been met, but who defines the level? Crick's belief in panspermia[25]—the idea that life on earth was seeded from space—was based on his assumption that insufficient time had elapsed for life to arise on its own after the planet became habitable, but most scientists have seen no reason to go that far. This is not an experiment anyone knows how to run, so which high priest should we believe?[26]

One possible response to the observation that the sciences (and certainly scientists) embody a variety of behaviors common to traditional religion is just to acknowledge that science is the religion of choice for some and be done with it. While I think this is the case, such an approach has limited usefulness. For one thing, it merely states the obvious. More importantly, it

---

22. Barfield, *Saving the Appearances*.

23. The role of electron orbits in scientific theory is addressed in an interesting encounter between Heisenberg and Einstein (Heisenberg, *Physics and Beyond*, 62–64). In grade school you may have learned about planetary models of atoms (i.e., where electrons orbit a comparatively massive nucleus) but things aren't quite that simple.

24. John 1:18a; 1 John 4:12a.

25. Crick, *Life Itself*.

26. Crick also wanted to hold religious beliefs to the same standards as manifest in the sciences (*Astonishing Hypothesis*), but which science should we chose? According to Bak, those with the most rigorous connection between theory and empirical evidence deal with relatively uncomplicated phenomena and those that try to explain truly complicated things (such as the economy) often fail miserably (Bak, *How Nature Works*; also see Taleb, *Black Swan*).

presupposes that a person must choose between science and religion just as she might opt for one traditional religion over another. Although there are people who embrace such an assumption, their reasons for doing so lack punch.[27] As an alternative, someone recognizing the power of both traditional religion and science might attempt to merge the two, perhaps along the lines of Stuart Kauffman's proposal that life, the biosphere, culture, economics, and the like are complex, creative, self-organizing, evolutionary systems—let's call that god.[28] At this point believing or disbelieving in God starts to look much the same but at least Kauffman is exploring, even if what he offers comes across as sophisticated pantheism.[29] A little sophistication never hurt anyone but, unfortunately, it seems to be in short supply in the views on religious faith extolled by many otherwise sophisticated people. Despite the fact that science and religion have co-existed for some time, they are still like bickering siblings who have not yet left their childhood. There are probably many reasons for this but building one's own religion—something even the supposedly non-religious do—is certainly one of them.

* * *

Characterizing science as a religion will, no doubt, irritate some scientists and possibly others who want to place as much distance between the two as possible.[30] There is no need to get carried away with the characterization, but those insulted by it might ponder whether it is because they believe I have misrepresented science or because I have upset their belief about science. Either way, that zeal to protect one's belief turf is part of what we have been considering. The scientist doth protest too much, methinks. In any

27. Doesn't it seem rather ironic (and not a little fascinating) that some of the scientists who are most outspoken about the flaws in traditional religion and are, consequently, quite adamant about segregating science and religion are the very ones who come across as most religious? It would appear that individuals overtly opposed to a merger of the two arenas have implicitly accomplished just that.

28. Kauffman, *Reinventing the Sacred*.

29. Pantheism is an easy response to the natural/supernatural issue but one that fails to satisfy many people. Theologians trying to reconcile God's immanence and transcendence have endeavored to replace pantheism with panentheism—the idea that God is not equivalent to nature but permeates it (cf. Peacocke, *Creation and the World of Science*). Although we might applaud the effort, it is difficult not to get hung up on semantic nuances.

30. I can imagine being accused of misunderstanding science or being too loose with my analogies, but as a scientist I know well how science works and how scientists operate. Although the science as religion simile proves nothing, it suggests much.

case, there is a simple test for anyone to determine if something (science, hobby, a person, etc.) has religious overtones for them and that is to evaluate the extent to which he or she praises it. Just because we praise something doesn't mean we worship it, but we will *always* praise what we do worship.[31]

---

31. There may be many reasons for criticizing various manifestations of religious faith but flippant derogatory dismissals of religious faith itself (as opposed to rejecting a specific instance of it) are no more valid than flagrantly embracing it simply because it is religious.

―――― CHAPTER 11 ――――

How Clear Is God?

There is no worse lie than a truth misunderstood by those who hear it ...[1]
—WILLIAM JAMES

Communication Difficulties

ONE OF THE MOST obvious factors distinguishing humans from other species is our marvelous faculty for language. There is good reason to believe that without it, we would have no culture of any consequence and would be little different from our primate relatives. Humans expect to be able to communicate with one another and we are frustrated when unable to understand or to make ourselves understood. Possible reasons for such failures vary widely and may involve a problem at the sending or receiving end of a message, perhaps due to a speech impediment or hearing difficulties (or poor handwriting or eyesight in the case of written language). Occasionally both ends are equally culpable, as when the parties speak different languages. At other times, the problem may be beyond the control of either sender or receiver, perhaps due to something as mundane as a poor telephone connection or ambient noise.

So, what about contact between humans and God? The theistic religions—unless they have evolved into a form of remote deism—assume some form of communication from (and perhaps with) deity but, if God is speaking to us, it is plain that different people are getting different messages. While some observe evidence for God everywhere they turn and claim to

1. James, *Varieties*, 205.

experience him in practically all aspects of their lives, others see and hear nothing. What are we to make of this? For Nietzsche, the answer was simple—God "seems incapable of clear communication."[2]—and I suspect that many will agree. Even those who deplore much of Nietzsche's thinking have probably wondered if God could have been a bit clearer on occasion. After all, how could God be clear if people are getting mixed messages?

The answer to that, of course, might depend on whether the problem is at his end or ours. Most of us have experienced the discomfort of sitting through lectures or trying to read books we could not understand, even though they were presented in our native language, because the subject matter was well beyond our reach at the time. Although we are sometimes prone to blame the speaker or author, most of the time we realize that the inadequacies are on our side. Is it possible that God seems unclear because we have not yet learned to understand what he is saying or how he is saying it? In fact, if there is a God, how could it be any other way? Why should people who complain that they can't understand algebra or Shakespeare or quantum mechanics or consciousness nevertheless expect to understand God instantly and consider it his fault if they can't?

I suppose that one response to this is that God should at least be able to say something fundamental and without ambiguity that we all could understand. Shouldn't the Almighty take the initiative and make it simple enough for us? But how simple should that be? Would God's message be deemed satisfactory if everyone was convinced of it? Presumably, we would want anything God says to also be of some consequence, yet it is all but impossible to think of anything of any import whatever about which everyone either is or could be convinced. Perhaps if God was trivial we could agree on what he says but then there would be no need to pay attention. In fact, the only kind of God people do agree on is the trivial kind—even within a specific religion people disagree about God in large part precisely because he is deemed non-trivial. While some people think that the multiplicity of views about God is reason enough to doubt his existence, disagreement is precisely something we might expect about a God of any substance. How could God not be complicated?

Nevertheless, we may feel justified in thinking that, because he is God, he could somehow span the gap between his complexity and our simplicity. Some religions do claim that God *has* made it simple enough for everyone—that he has communicated in plain language—and that the problem is not on his end. If that is the case, it may be prudent to consider what problems reside with us that inhibit communication. We can begin by asking what we expect to find.

2. Nietzsche, *Beyond Good and Evil*, 66.

On Touchy-Feely Gods

Several years ago, after my oldest daughter had lost one of her baby teeth, I found the following note accompanying her solicitation and gratitude for the funds customary on such occasions: "Dear tooth fairy, thank you. Please give me a picture of you. Put signal here." What she expected to receive in the way of an image, I do not know, but I am inclined to believe that a photograph or drawing depicting one of the fairies inhabiting a storybook or movie would have satisfied her quite well. I am also relatively certain that a picture of me or her mother would have been a big let-down and perhaps even considered a bad joke.

While we smile at the naïveté of a child, we each leave our own notes for God asking for a picture or sign, oftentimes with little thought as to what we actually expect to receive.[3] But what are we looking for and how would we recognize it, even if it was (or has been) given? While it may seem reasonable to expect a signal we cannot miss, we must also ask if such a signal is even possible.

Rather than consider this in terms of generalities, however, let's try another experiment. Your task is to imagine that you are hoping for a picture of God that will help you to see him clearly and, in light of that hope, to ask yourself what you might reasonably expect to see that would satisfy your desire. Certainly no photograph would do—most people quickly grow out of their old-man-with-a-beard image of God—but if not that, what? Through the ages various idols and icons have come and gone but it is improbable that any physical representation will prove satisfactory, which is probably why some theistic religions tend to portray God as spirit. Naturally, it is difficult to photograph or draw a spirit but that doesn't mean we find the concept meaningless—quite the contrary. We regularly use the term to represent the deep essence of a person, team, or nation that is otherwise beyond our ability to adequately describe. However, it should be obvious that we can also use it as a stand-in for the products of our imaginations. Consequently, I am not suggesting that one leap to the conclusion that the concept of God as spirit, based on our inability to picture a material God, somehow equals proof that God exists. By that line of reasoning we might believe anything. What I am proposing is that God's supposed failure to give us the picture we think we deserve might possibly be due to a faulty assumption on our part that such a picture is possible.

Although it is apparent that not being able to see God is what we would expect if he does not exist, it is also precisely what we should anticipate if he

3. Anyone picturing God as little more than a glorified fairy is going to have trouble understanding my comments in this section.

is as powerful as the major theistic religions have portrayed him. The image of any God whose attributes and abilities exceed our own by an amount sufficient to encourage us to call him God would necessarily require a tremendous reduction in scope to be perceivable by us. Anything we are capable of conceiving about such a God would (and does) in some sense bring him down to our level, which means that he would quite possibly appear unlike anything we have imagined[4] and that such an appearance could at best be a partial representation of his total being. If there is a God and the photograph is blurred, it is hardly surprising.

What Kind of Evidence Will Do?

Even if we can get comfortable with the idea that a complete image of God is beyond our ability to comprehend, that doesn't prevent us from expecting, and I think rightly so, some kind of evidence that we can process. But what evidence should that be? It may be difficult to acknowledge that any evidence we hope to understand pertaining to a boundless God must have bounds but a greater problem—perhaps the greatest problem—is to determine what evidence would be convincing. The real question we must all face is this: What would it take to persuade us that there is (or is not) a God?

In the movie *Oh, God!*,[5] George Burns, playing God, appears to a skeptical John Denver, who naturally wants proof that he is really in the presence of deity. Denver's request for a change in weather is quickly granted by a downpour within the car in which they are riding. I don't know about you, but I'm relatively sure that I would have asked for something else besides rain. But what might that be? Imagine that someone appears before you and claims to be God. What evidence would you consider adequate to support the claim?[6]

Some people, such as author Michael Shermer, believe that it is not "possible to know if there is a God or not."[7] I hope you'll ponder that perspective long and hard because it may reveal why many people—atheists and theists alike—are so adamant about their positions but frequently come

---

4. Although it would seem that, by necessity, he would look like some of the other things we are capable of perceiving.

5. Reiner, *Oh, God!*

6. Richard Dawkins noted during an interview on the National Public Radio *Fresh Air* program (March 27, 2007) that he looks to the evidence and sees none for God. Our interest here is what kind of evidence it would take for him to change his mind (or for his theistic polar opposite to change his). What kind of evidence is he looking for and why should that be the preferred kind?

7. Shermer, *Science of Good and Evil*, 5.

across as missing something quite fundamental in their rationale. To digest this, it is not necessary to reject the statement entirely. In fact, I think we can agree that it may even be half right—if there is no God then we could probably not expect to be able to know it. But the possibility of knowing that there is a God is a different issue and not just a mirror image of the other. For example, it seems impossible to know that someone will never run a three and a half minute mile, climb Mount Everest in gym clothes, or produce a grand unified theory in physics. Although one may have what she considers good evidence supporting or refuting her belief in the occurrence of any one of those things, that evidence could never be enough to remove the possibility that someone may just do it, in which case the issue becomes moot. Similarly, a person may believe he has considerable evidence for the non-existence of God but there is always the chance that he will run into God tomorrow.

On the other hand, if there is a God, the possibility of knowing that he exists is probably much more likely than knowing that he does not unless (1) he doesn't want us to know or (2) we don't want to know. I will ignore the first case for the simple reason that, if it is true, there would be no way to distinguish it from the situation in which there is no God. That leaves us with the prospect that no amount of evidence might be sufficient to convince us that God exists. This, I think, is the real sticking point inhibiting the development of rational faith pertaining to matters concerning God but it would be wrong to think that the root problem is restricted to atheists. The theist for whom no amount of evidence would be sufficient to convince him that God most likely does not exist or that God's attributes or behaviors are not what he has imagined them to be is similarly debilitated.[8]

Clearly this is not an attitude of which to be proud—it suggests a certain closed-mindedness which the atheist habitually criticizes in the theist and vice versa—but the way to demonstrate that it does not characterize our own viewpoint is to be able to say what we would consider to be adequate evidence for or against God. If some evidence could, in principle, be satisfactory, then what would it be? Please don't say you'll know it when you see it. That may be the case but such a response is probably more likely to reflect a failure to consider the possibility that no amount of evidence could be enough.[9]

---

8. In other words, mounting evidence against God's existence can be denied with obstinacy no less dogmatic than that shown by one who ignores increasing evidence for his existence.

9. It may not be the same to say that nothing would convince us as to say that we don't believe there will be any evidence that could convince us but from a practical standpoint the results are likely to be the same.

When Jesus remarked that, "Unless you people see signs and wonders, you will never believe,"[10] it almost sounds like an acknowledgement that people would yield to some standard of evidence. Certainly some will and that is what we should hope, but such a conclusion does not logically follow from what he said. Let's make this clear by rewriting his statement in "if-then" form: "If you do *not* see signs and wonders then you will *not* believe." Note that this is not equivalent to saying, "If you *do* see signs and wonders then you *will* believe. Logicians call the latter the inverse (of the original) and it is a logical fallacy to infer it (from the original). Clearly, we don't need a logician to tell us what we have observed from encounters with any number of irreconcilably stubborn people. In one of his parables, in fact, Jesus made his conception plain regarding how some people will handle evidence: "they will not be convinced even if someone rises from the dead."[11]

It is one thing to say we don't believe that someone rose from the dead (presumably because we find the evidence inadequate) but quite another to be in a position in which nothing could make us believe it. It is different still to admit that, even if we did believe, it would not be enough to convince us of the existence or power of God. I want to make it clear that the latter response is not necessarily a logical error—one may think that rising from the dead can be attributed to factors other than God, although that would probably satisfy most people. Some, however, may have previously decided there is no God and they must look for alternative explanations. Even that perspective is understandable to a point, but only if the person holding it will say what, if not rising from the dead, they would accept as evidence for God. If they cannot or will not, it might be prudent to be wary of anything else they have to say about God's absence. In order to avoid hypocrisy, of course, Christians who might have agreed with my assessment to this point must be willing to say what would make them willing to disbelieve that Jesus rose from the dead. If there is nothing that could do so, it is hard to see how one is to distinguish the closed-minded theist from the closed-minded atheist with respect to the foundations for their respective faiths.

I want to pause here and emphasize that, although I am fond of symbols, I hope to be taken quite literally on these points. It is not unusual to hear the non-religious lamenting a perceived lack of rigor among the religious in their approach to evidence,[12] but from what I have seen, the problem has less to do with one's religious beliefs and more to do with how one approaches the very concept of evidence in the first place. If someone

---

10. John 4:48.
11. Luke 16:31.
12. Cf. Crick, *Astonishing Hypothesis*.

thinks religion should be subject to the same level of confirmation as the sciences—and I have already suggested that, with respect to matters so potentially important, religious folks should welcome every opportunity to refine their beliefs—then they should be prepared to specify the experiments that would falsify or support their hypotheses. But, because every hypothesis is built around assumptions, one must acknowledge from the outset that the success or failure of any particular experiment may tell us as much about our assumptions as it does about God. For example, the kinds of tests one typically hears proposed may involve requests for a particular sign (along the lines of John Denver's character's request for God to manipulate the weather) or attempts to quantify the efficacy of prayer. I expect the astrophysicist would suggest something he considers a bit more impressive than a violation of the weatherman's prediction—perhaps the reversal of planetary orbits would do—but I have no intention of trying to enumerate all of the similar requests that might be forthcoming. Note, however, that any hypothesis positing that a positive response to such requests would count as evidence for God and that the absence of a response would work against it is based on the assumption that God is some kind of cosmic bellhop just waiting to service our whimsical desires with the kind of evidence we demand (as opposed, perhaps, to someone who is hoping we might re-evaluate the evidence already provided).[13] If you think about it, the God discredited by such an approach is probably not worth writing home about anyway. So, while providing a sound foundation for one's theistic or atheistic beliefs is a noble goal (and one I have endorsed throughout this book), it is difficult to be patient with people who clamor for a scientific approach but neglect (or refuse) to elaborate their assumptions. My fellow scientists, in particular, should know better.

If it is unwise to assume that a God worth his salt will respond to any old request for a sign, what are we to do? For starters, we must be careful that we are not using this as an excuse to defend a God who isn't really there. God's refusal to answer the call of every solicitor doesn't mean he is not at home but it might. Evidence is crucial but it is probably worth remembering that, if there is a God, he was here first. Consequently, in a very real sense, he is not the object of our experiment but we of his. It might, therefore, be more reasonable to expect that, if he wanted to be known by us, he would provide the evidence he deemed appropriate as opposed to jumping at our

---

13. Some of our proposed tests (i.e., requests for evidence) must look awfully ridiculous to God but even a fanciful request seems better than none at all. At least it is a start. If our expectations remain on that level, however, we may never progress to the point of learning much. That God would, perhaps, on occasion humor the sincere seems possible.

beck and call. Rather than looking for God to perform a one-time trick to convince us, we might try accumulating evidence through history. This is what the major religions have done. Anyone genuinely interested in exploring what can be known about God, then, would do well to consider the hypotheses and evidence provided by those various perspectives.

Although it may seem that this is what most people do, I don't think that is really the case. Because religious beliefs tend to be passed along much like genetic information and to exert a comparable influence, it is possible for the possessor of those beliefs—it might be more accurate to say the person possessed by them—to give little or no thought to the relationship between what their inherited religion (be it atheism or theism) hypothesizes and how the evidence supports those hypotheses. In fact, it is likely that many people never see their religious beliefs in such terms. If we do, however, we must confront the possibility that our hypotheses may require modification or even replacement—this is part of the price we pay for attempting to put them on a sound foundation. That means we may have to change our opinion regarding the veracity of some event or what it means, about how God works, and even whether there is a God. But if we expect to evaluate our hypotheses honestly, we have to be willing and able to determine what evidence would be convincing.

Religious Hypotheses and God's Clarity

If we expect our understanding of religious faith to be thorough, we must allow for the possibility that it is our own religious hypotheses that contribute to perceptions that God is unclear. Consequently, I am going to use a structure called a decision tree to help conceptualize the role of hypothesis formation and evaluation and to support the process of thinking more deeply about the issue of God's clarity. Figure 1 shows an example that indicates the types of hypotheses one might adopt with respect to religious claims (Because the decision being made at each node in the tree is really about a hypothesis, we'll be a bit more specific and call it a hypothesis tree). The hypotheses are arranged in hierarchical fashion to reflect the fact that some are invoked only in the presence of others. To actually use the tree one must apply it to a specific religious claim, preferably one of the more prominent ones traditionally promoted as evidence for some particular religion. This might be a claim about origins, miracles, revelation, changes to human lives, and so forth. The tree makes it plain that hypothesis formation (and hence the assumptions on which hypotheses are based) occurs at a variety of interdependent levels. There is no such thing as a simple, stand-alone hypothesis when it comes to major religious beliefs.

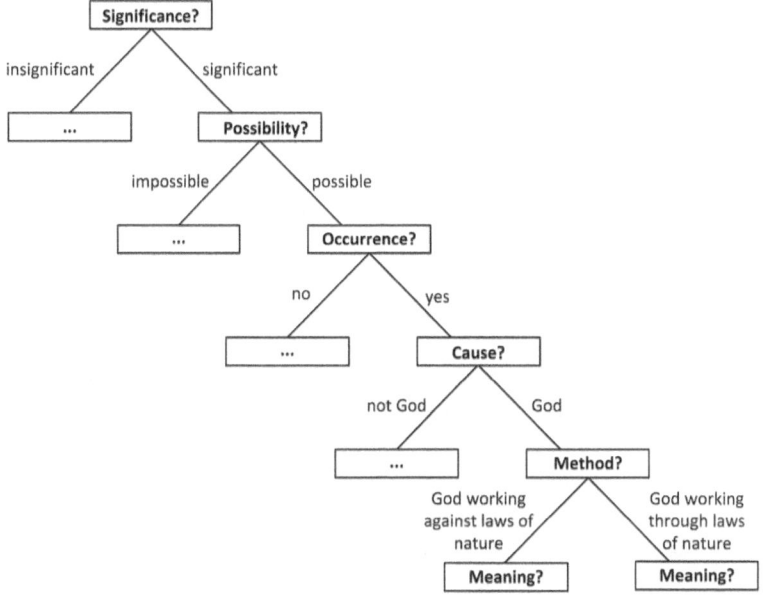

Figure 1: *Hypothesis tree for religious claims*

Is a Religious Claim Significant?

As you can see, just getting started is not without its difficulties. Even the issue as to whether some claim is significant to a religion or not is a hypothesis. Perhaps all followers of the religion accept it but, more likely, its perceived significance depends upon whom you ask. Because the claim should be specific, the context in which it is presented will be crucial with respect to hypothesizing about its importance. The scriptures of a religion, for example, may report many births and deaths with only some appearing to warrant special status. Even claims about someone coming back to life after being dead may be deemed relatively insignificant if people are routinely raised from the dead in the religion under consideration or if the event looks like an instance of cardio-pulmonary resuscitation. In any case, believing that a religious claim is insignificant will incline one not to give it much more thought and the rest of the tree will remain unexplored. On the other hand, incorrectly hypothesizing that something is significant can result in wasting a good bit of time that could be better spent on truly important issues. Because a decision whether the claim is true or not is only made later, one can still have some belief at this point about what the majority of devotees of the religion consider important (regardless of what is personally thought).

Is the Claim Possible?

A claim that possesses significance for a religion may be about something whose likelihood is not obvious. Mundane events garner little attention, so the next hypothesis faced involves what is believed about the possibility that the object of the claim is true. Is it possible that the Israelites were miraculously delivered from bondage in Egypt, that Mohammed received a unique message, or that Jesus fed thousands? This is not asking if we have ever observed something on the order of the claim under consideration or know someone who claims to have done so (although either would be helpful), but whether we believe it could have happened (or could happen). As at other levels in the tree, any hypothesis here can be nuanced. However, deciding that something is impossible is a show stopper, shutting the door to further exploration of the claim under consideration and, possibly, religious claims in general. It is almost always possible to get to this point in the tree but some of us will go no further.

I imagine it was significant religious claims Mencken primarily had in mind when he referred to "belief in the occurrence of the improbable"[14] but, to his credit, he didn't say "impossible." We stick our necks out a long way if we go so far as to declare something impossible, particularly in a universe where quantum physics suggests there is always some probability for anything. Most people, if pressed, would probably allow for the remote possibility of some religious claim but, if the possibility is believed to be too far-flung, it will be treated as impossible and search of the hypothesis tree will terminate at this point. Of course persons hypothesizing that some religious claim is too impossible to warrant further consideration may really be saying that no amount of evidence could convince them. In that case they will never get to the next level of the tree where any actual evidence must be considered. Such a mindset (really, a set mind) defines away both the event and any God who might be behind it.[15] No wonder God might

---

14. Mencken, *Prejudices*, 267.

15. It seems odd that one could unabashedly believe that anything is possible in the quantum world with some probability but state that some things are impossible in the religious world under any circumstances.

appear unclear in such a case! This—the epitome of the closed mind—is ironic because such people are frequently the first to condemn the "rigidly held" beliefs of others despite the fact that those alternate beliefs could only be obtained by opening this door.

Might a Claim be True?

If one does acknowledge the possibility of some religious claim, the next step is to question its veracity. In the astronomically large space of possible events, relatively few actually occur. Thus, although evidence is important to hypothesis formation at each node in the tree, I think nowhere is it more critical than here. Most open-minded and curious people will make it at least this far, but the hypotheses formed at this point with respect to the major claims of a religion ultimately express one's beliefs about the overall truth of that religion regardless of what is thought about God. We may acknowledge the possibility of a miraculous parting of the Red Sea, a special revelation to Mohammed, or a physically inexplicable multiplication of food but deny the reality of all of them. On the other hand, if we hypothesize that something is almost impossible but find the evidence for its occurrence convincing, the very near impossibility will make it seem all the more special.[16]

In any case, this is not the place for wishful thinking.[17] In fact, we can gain special insight at this point into our entire religious belief system by considering the question raised earlier regarding what evidence would be sufficient to convince us to believe the opposite from whatever it is we currently believe. If, for example, we discover that there is really no evidence that could convince us that some hypothesized event occurred, we have come too far down the tree and should back up to the previous node, admit that our real hypothesis there is that the claim is impossible, and face the likelihood of our own closed-mindedness. On the other hand, if no evidence could convince us that an event did not occur, then we should ponder

16. This sounds like C. S. Lewis remark that, "Reality, in fact, is usually something you could not have guessed. That is one of the reasons I believe in Christianity. It is a religion you could not have guessed" (*Mere Christianity*, 36). Of course, it is not hard to imagine someone from another religion making a similar claim. Any guess, however, is only possible within some context and the better the context, the better chance there is for the guess to be correct. However, even supposedly rare events (i.e., those that seem almost impossible) in one context may appear inevitable in another. It may only be our lack of access to the full context that makes an event seem rare.

17. Religious belief is often claimed to be the product of wishful thinking but desire cuts both ways—someone might not believe in God because he doesn't want to. Wanting there to be a God or not will color the ability to apprehend any that might exist but that doesn't mean one has to be impervious to the evidence.

our disdain for evidence and wonder why anyone else should accept our point of view.

Some people may get to this level in the tree by only imagining that they are open to evidence. However, if they really think the event impossible (or assume it to have such a low probability as to be so) then, even if evidence did actually exist (perhaps for a radical claim like someone returning from the dead), they would be forced to discount it or to postulate some alternative explanation (it must be a clone, I was hallucinating, the person wasn't really dead, etc.). Note that this would be the case even if they observed the event firsthand! Any smugness about not being in that group should be tempered with the knowledge that the best hope for observing whatever evidence might exist—much less interpreting it correctly—is to recognize the biasing effects of previous assumptions and hypotheses which frame our beliefs. If we realize that an assumption about the impossibility of something can keep us from seeing the evidence then maybe, just maybe, we can process that evidence.

This is one reason it is so important to ask what evidence would have to exist to convince us for or against a religious claim. That question may boil down to determining what kind of evidence we would consider adequate and how much of it we might require. If a claim is about a historic, one-time event, it would be silly to expect to be able to duplicate it. That leaves us to evaluate the historic record: Are there enough eyewitnesses? Are they credible? How do we know? How were their accounts transmitted? Do other claims in the religion support or refute the claim under consideration? How? Could someone have ulterior motives for providing false information? What might those motives be and what evidence is there for it? Are there alternative explanations that make better sense? Why? How have our preconceptions influenced our answers (and questions)?

For such questions to be meaningful they must be directed at specific religious claims but, for maximum benefit, they must also actually be answered. Are two witnesses inadequate but four enough? If not four, how about 400? Must the accounts be written down by the person who saw them or could they be dictated to someone else to record? How many links in a chain of verbal transmission are permissible before you begin to question the credibility of the report? Would you refuse a manuscript but accept a video?

Responding to such questions can help us understand why we believe or disbelieve a particular religious claim but I want to suggest that, at least conceptually, we can go even further. Let's imagine that there is someone—we'll call him Jules—who is one of those folks who has decided regarding some religious claim that, not only is there insufficient evidence to convince him of its truth, any number of additional historic accounts would also have no impact on his hypotheses about it. This is a bit narrow but we'll assume that Jules has his reasons. Because we are only imagining this, we might as well go ahead and pretend that Jules has access to a time machine that would allow him to return to the historic setting for the event under consideration.[18] What would Jules need to see in order to change his hypothesis?

If Jules attempts to answer this question it will tell him a great deal about himself. This includes the fact that, if there is a God, then any hypothesis about what counts as acceptable evidence is also a hypothesis about what God should have done.[19] Yet if we ever discover that our approach to religious claims is really characterized by an attitude in which, unless we had been there, there is no way we would believe (or disbelieve) them, we are actually professing a kind of impossibility for the objects of those claims.

Was God Involved?

It is unlikely that anyone could find reasonable evidence to support all religious claims (particularly since many are quite contradictory) so it is probably apparent that some claims must be discounted. However, it is also reasonable to assume that at least a few (perhaps many) will be deemed truthful, and not just by the devout. One doesn't have to be a Muslim to believe that there was a real Mohammed or a Buddhist to believe that Gautama Buddha actually existed. We may doubt other claims associated with them but even atheists acknowledge the lives of the chief figures associated with many religions. Consequently, with respect to at least some religious claims, we are likely to find ourselves at a point in the tree where we must hypothesize about the cause of something we have decided is likely true. Now we must consider whether there is reason to believe that God is or was involved in the claim we are confronting. While a hypothesis at the previous node might open the door to accepting a religious belief (and ultimately a religion), what one hypothesizes regarding cause determines whether one walks through that door or not.[20]

Here, evidence may be harder to recognize, particularly with respect to what it signifies. A person may comfortably accept the occurrence of an

---

18. Previously, I suggested that it is silly to expect to be able to duplicate a one-time historic event but thought experiments don't carry the baggage that reality does.

19. If there is a God, one might be wary about setting the ground rules for him.

20. Earlier we considered Jesus' comment that, "Unless you people see signs and wonders, you will never believe" (John 4:48), but the record shows that some people, even when they saw them, only believed the signs and wonders, not what they supposedly signified (Cf. John 11:1–53; Mark 3:1–6). This is the difference in sight and insight.

event without seeing any reason to attribute it to God (or not to, for that matter). So, what kind of evidence would it take to think that God was or was not involved? Perhaps the most obvious answer is something that is otherwise hard to explain—something that would seem to mandate God's involvement. But we've already seen how easily this can lead to a "God-of-the-gaps." Clearly we should not allow that to prevent the introduction into evidence of something amazing (or else there could be no possibility of believing anything amazing), but it does mean we must be wary of the implications of claiming that because something is currently without other explanation it must be due to God. Of course, we've already considered the possibility that, even if something has an explanation in terms of physical, chemical, biological, or psychological theory it may, nonetheless, still be due to God (although it may require us to modify how we perceive him).

In any case, while a single momentous event may appear to provide irrefutable evidence on which to hypothesize God's involvement, a conjunction of evidence-based claims seems even better, avoiding as it does the need to stake one's entire belief system on a single assertion.[21] Ideally we might look for a succession of events that not only suggests God's involvement because of their unusual nature but that also complement one another, standing together in a sort of logical harmony, perhaps demonstrating or at least suggesting some underlying meaning or purpose (presumably that suggested by the religion under consideration). The events identified in such claims should be foundational and not peripheral to the core message of the religion and should not unduly confine God (just yet) to some specific mode of operation. (That can be done further down the tree.) Here we can begin to see that such an approach may necessitate revisiting some of the prior nodes in the tree and might make us hesitant to assign confidence values to any particular event or claim at this node until we have considered some of the related lower-level hypotheses. A real explorer will move freely

---

21. One might be inclined to look for events and, particularly, conjunctions of events that seem unlikely to have occurred entirely by chance but that approach must be based on the hypothesis that God has nothing to do with chance (cf. Bartholomew, *God of Chance*). A person who hypothesizes that God works through chance will find it difficult to then turn around and use a low probability event as his basis for evidence because there would be nothing to distinguish it from any other random event. On the other hand, it is problematic to think that we can count a string of answered prayers pertaining to low probability events as evidence for God's existence or action. We've already discussed the problem with assuming God will answer any old prayer but even if we can point to ten or twenty in which we detect unequivocal evidence of his involvement we must be careful that we are not disregarding as many others in which we see no evidence that the prayer got past the ceiling (Bacon, *Great Instauration*, "Novum Organum," aphorism XLVI).

between nodes at various levels and will seriously contemplate the potential relationships between hypothesis trees for different claims.[22]

As this is done, many factors are at work involving the potential for multiple claims to support or contradict each other. In particular, the probabilities assigned to individual claims can be used to strengthen or weaken the basis of support for a system of linked claims.[23] Knowing this could encourage a wishful thinker to see conjunctions where none exist, but it might be more disastrous to miss those that actually do.

It is hard not to think that any religion that is true (or mostly true) should exhibit some emergent property that transcends any individual event of that religion. I am hesitant to specify what a religion should or should not do but, just as the wonders of consciousness emerge from a collection of non-conscious components in the brain, so it seems to me that we should not be surprised to find that a genuine religion, if it exists, yields something equally dramatic but less clearly revealed in its parts.

If this is so, we would expect that the scriptures of a religion, for instance, should paint a coherent picture of God. Yet as we've noted, the hypotheses one adopts about the origin of those scriptures can make a huge difference in what one sees. For example, one possible reading of select portions of the biblical Old Testament—based on the idea that those passages were given by God on a silver platter—hypothesizes that God ordered extermination of non-Israelite peoples in apparent support of the idea that the Israelites were favored. There is a certain consistency there but it leaves the impression of a rather provincial sort of God. An alternative hypothesis suggests that those passages reflect a growing but rather immature view of God that found a different expression in later Old Testament voices such as Isaiah who saw the Israelites as "a light for the Gentiles,"[24] a theme echoed by Jesus who described both himself and his followers as "the light of the world."[25] Here, there is the emergent theme of God's universal love that draws one's attention and suggests further scrutiny. Regardless what we believe about these specific examples, should we let a history of limited views of God be taken as evidence against him?

---

22. By this time the exaggerated simplicity of our hypothesis tree should be evident. In reality it should be much more like (what a mathematician would call) a graph, with arcs (edges) extending between nodes (vertices) in all directions and with a graph for any specific claim linked to graphs for many others.

23. Statistically speaking, we would be interested in "conditional probabilities."

24. Isa 49:6.

25. John 8:12; 9:5; Matt 5:14.

What Method Did God Use?

We have been considering how one might look for evidence pertaining to God's possible involvement in an event (or collection of events)—evidence that would contribute to a clearer picture of God—but, even if we do hypothesize his involvement, we are still faced with the form that involvement might take. Does God work through known scientific laws (i.e., can hypotheses about God and science be compatible) or does he suspend them for special purposes (i.e., are science and God mutually exclusive)? Or, are there physical laws within which he operates that we have yet to uncover? Perceptions regarding origins, miracles, answers to prayer, and the day-to-day course of human lives all reflect various hypotheses about how God interacts with the universe we know.

For example, consider possible hypotheses about how God effects change in human behavior. As we saw earlier, everything we know about behavior suggests that it is the product of brains. Using the binary alternatives shown in the tree, then, one might postulate that, with little concern for the laws of physics, God makes a sudden change to a person's brain leading to a change in behavior. Alternatively, someone else might suggest that the process of reading and thinking about God, coupled with other actions, creates a neural landscape that engenders a change in behavior. The first hypothesis certainly gives the impression of being easier to understand but is quite problematic for anyone familiar with how our bodies function. This is interesting because it suggests that simpler equals clearer, but in this case the less clear explanation may be the better one. In other words, God may not always be as clear as we would like because the phenomena with which he is involved are complex. Every student of a difficult subject understands this—we would like a simple solution but not at the expense of accuracy or explanatory power. Besides, serious students know that difficulty diminishes with study and that subjects that at first look terribly opaque can become significantly clearer with time.

One might think that people who have agreed on the same basic hypotheses up to this point would be amicable, even if they do not see eye to eye with respect to how God might be involved in the world—at least they agree on God! Sadly, that is not always the case. The resulting turf wars may reflect the perception that a change in hypothesis about method might undermine the entire edifice on which a religion is thought to be constructed—certainly there is the potential for radical change in how one understands various components of the religion.[26] The irony here is hard

---

26. Note the potential for feedback in this process. For example, a change in hypothesis about how God is involved in the world can change how one reads and understands

to miss. If the prevailing hypothesis turns out to be inferior to another, it is possible that its supporters, in trying to defend their particular point of view, are actually the very ones muddying the proverbial waters. It is not God, then, who is unclear but those who think they are protecting him (possibly in the name of clarity, no less)! As a corollary, the most trustworthy advocates of any theistic religion (that is supposed to reflect the truth about God) will necessarily be explorers and not those who protect at all costs their hypotheses about him.

Clearing Things Up

Some years ago I was scuba diving in the Gulf of Mexico in water with supposedly good visibility but was finding it nothing special. As I turned my head, however, the small amount of water that inevitably collects in a diver's mask flowed over the face plate and dissipated the fog that had, unknowingly to me, accumulated there. At that instant the ocean opened before me but that underwater vista was not all that became clear—so did the knowledge that the visibility had been fine all along and that the problem was with me.

This is only a metaphor, of course, and perhaps God really is unclear. But any honest approach to this issue must also acknowledge that the problem may be of our own making. Each of us forms hypotheses about the religious claims we hear but, most of the time, we do so haphazardly, primarily as a reflex action. Consequently, I have proposed using a hypothesis tree as a way to help formalize thinking and expose any flaws therein—several of which are especially pernicious. The person who seldom pauses to stop at each node and evaluate the basis and ramifications of their hypotheses is susceptible to implicitly sliding down the right side of the tree pell-mell and anything will be believed. Conversely, someone who has precluded the possibility of certain events is likely to truncate their deliberations prematurely and nothing can be believed. One sees things that aren't there while the other fails to see things that are. For the former everything seems clear while for the latter nothing does, but both may be in the dark, partially because they have neglected to establish any real criteria for religious faith.

The hypothesis tree is a way to do just that but only when its branches are carefully explored. If there is no God, that might provide the reasons for saying so but, if there is, this is likely to be the most important tree one could plant.[27] If the tree strikes you as overly-simplistic, I encourage you

---

the scriptures of a particular religion but it is also true that a careful study of those scriptures can lead to a change in hypothesis about how he operates.

27. Anyone getting far enough into the process must ultimately ponder the meaning

to develop your own. Surely some of the hypotheses one must make are not binary—in fact, all are probabilistic—but confronting them is the road to clarity.[28]

Warning, Limited Sight Distance (On Faith and God's Clarity)

Giving God the benefit of the doubt about his role in the clarity issue and accepting responsibility that any perceived problems might reside with us is a tough diagnosis—one, I suspect, that few will be eager to accept outright. But we must have our scapegoat so, if it is not God or us, then why not place the blame on faith itself? In other words, it is not hard to imagine that we must confront those unfortunate circumstances and times when God seems unclear with faith but that, for those situations when God is clear, we can blithely disregard it. It is not at all uncommon to hear this from atheists in their attempts to disparage both God and faith but it is just as common to detect it among theists who will be grateful when the regrettable need for faith finally goes away. I called attention to these ill-posed perspectives in the opening chapter and, throughout the book, have been constructing the weaponry with which to blow this nonsense out of the water. Here I would like to bury the corpse.

To think in any way that faith is the problem or that, more specifically, faith is a religious problem is to entirely misunderstand both faith and God. Both theists and atheists alike can begin to see the absurdity of such a position by simply reframing the content of this chapter as a simple question—"How clear would God have to be so that we don't need faith?" The straightforward answer is, "He can never be *that* clear."

God will never be clear enough to satisfy us at any given moment nor can he. This is the price we pay for not being God ourselves. Consequently, there is no other way to know about God or to know him except via probabilistic assessment of and response to the evidence we have for him—in

---

of their hypotheses—most religions, after all, are not meant to be some abstract intellectual exercise. I have not included a single node for meaning in the tree because the issue is really relevant throughout.

28. Those struggling to find merit in religious claims about God might do well to try to fabricate some claims of their own, not out of an overblown sense of self-importance but to more clearly understand the claims that are already out there. Ultimately, it is critical to test existing evidence lest we end up making God what we want or think he ought to be. Of course some people will always believe that is what has already been done.

short, by faith.[29] As I've indicated earlier, we get a glimpse of this in our relationships with people (whom we are also unable to fully know) in the sense that we can never be entirely sure of their motivations, intentions, or feelings. Faith fills the gaps in our contacts with others in terms of what we believe about them, think they will say, imagine how they will behave, and so forth, but when the infinite touches the finite the gap is indescribably larger. Anything God does to bridge it will necessarily require faith on our part. It could be no other way.[30] Furthermore, although the theist can live with the faith that some things will become clearer, there is no point of ultimate clarity (except for God himself). We've already seen that it won't happen in heaven—how could it?[31]

The history of religion and personal experience as testified by many individuals suggests that God becomes clearer with both time and effort. If there is a God, this may not only be reasonable but inevitable. Consequently, if this premise is correct—that religious understanding of God has been and continues to be a matter of progression, potentially mirrored in the growth of insight at the level of each individual (and, in fact, driven by such growth)— then we might wonder if there is something special about the process of discovery itself. I think there is and want to suggest several reasons why.

Most obviously, this is how the universe works. In every other area of life, anything of significance is learned through a process of discovery. Yet, when it comes to knowledge of God, some individuals apparently feel that things should be different. People who spend years of serious effort acquiring advanced degrees in order to make difficult subjects in their chosen field comprehensible are often the first to complain that God (if he exists) should be clearer. Sometimes these are the same individuals who discount miracles, which is highly ironic because that is precisely what they are really asking for here. The idea that God should gift-wrap his evidence and send it by express mail is not limited to atheists, however. People who have grown overly comfortable with the concept of the miraculous are apt to expect that as the only way knowledge of God will be divulged and to assume that they need merely sit back and wait for it. As we've already noted, it is not necessarily

---

29. Objections that we could also know God via revelation fails to recall that revelation is just another possible form of evidence.

30. If there is a God and this is true, some potentially legitimate hypotheses are: God has made himself known as clearly as we can handle; there is enough evidence already; no additional evidence could be any more convincing. Alternatively, one might wish to postulate that God might continue to supply new forms of evidence.

31. People who take Paul's words in 1 Cor 13:12 too literally will disagree but, if the denizens of heaven have the kind of understanding those folks imagine, the envisioned insight must be equal to God's and that raises an even bigger problem. Some, of course, imagine they will be gods.

clear that something is a miracle (and if one sees a miracle in everything then any characteristic you like can be attributed to God). Thus, neither atheist nor theist should be surprised that much of what is substantive that might be learned about God will involve a process requiring significant effort not entirely unlike that to which they are accustomed in other areas of their lives. Sleep learning only occurs in fiction.[32]

Less obvious, but equally (if not more) important, is the fact that discovery is integral to our freedom. In the immortal sentiments of Fleetwood Mac, "You can go your own way."[33] How could we be free to choose unless we had to process the evidence? How could we be free if we were coerced into this or that belief? When people force themselves on us we consider it an intrusion. Are complaints about God's lack of clarity actually a request for him to make it impossible for us to resist? Is that what people really want?[34] In human relationships we properly reveal ourselves to others because there is a desire on their part to know us. If we do it under other circumstances it is classified an invasion of privacy or worse. God, however, doesn't coerce us.[35]

This, clearly, is only a hypothesis. Yet, if there is anything to it, our engagement in the process of discovery might say something potentially significant about our relationship with God, beginning with the extent of our desire (or not) for such a relationship. Some time back I was in the process of making plans to take a climbing trip with my two sons. I had laid out an itinerary that promised to be both challenging and memorable, but shortly before we were to leave I was informed that they preferred to stay home. Other attractions—yes, a girl was involved—were just too much competition for dear old dad. I could have pressured them, but what good would that do? If they had gone under duress, how could our relationship flourish? What quality would the memories have? In order to maximize the chances for a successful trip, it was important that they want to go. This, obviously, is the nature of a relationship. In order for it to work, both parties must want it.

---

32. This is not to deny that God may make some things so plain as to be accessible to anyone. But neither does it mean that because something is plain it will be accepted or its significance comprehended.

33. Buckingham, "Go Your Own Way."

34. Well, want it or not, this is apparently what some people do believe. Both traditional scientific determinism and certain religious perspectives posit a world in which some people cannot help believing in God while others could not if they wanted to. On the other side, it is interesting to contemplate whether the need for discovery (and even the perception that God is unclear) can be taken as evidence for our freedom.

35. It is not necessary to conceive of God as some inexorable black hole from which escape is impossible and this is one reason why.

All of us have learned that, whether it is dating, marriage, or just general friendship, it's no good being with someone who doesn't want to be with you. Communication will be trivial, meaning superficial, trust shallow, and joy non-existent. Relationships with imaginary people can go any way we want them to because we can project their response and there can be no resistance. But with real individuals there are criteria for establishing and maintaining any possible relationship and the first is a desire for it. If God is real, perhaps he only reveals himself to those with whom he anticipates the possibility of a good relationship—to those who want to know him. If that is the case, how else would we indicate our desire than by seeking for him?

Upon returning from my (solo) climbing trip, I was able to tell my boys about the adventure, although in no way was it as understandable to them as it would have been had they been with me. In fact, any complaint that my description was not entirely clear would appear ridiculous because they realized they could have been there. Naturally, they knew better. Some people, on the other hand, show little hesitation to complain about God's lack of clarity even after choosing not to travel with him, their lack of desire to know him or know about him evidenced by the simple fact that they are not seeking him.

The final reason for suggesting that the process of discovery may be special is that in a very real sense it levels the playing field for those who are seeking God. If some bizarre mathematical-theological hybrid suddenly appeared on the scene with an unassailable mathematical proof for the existence of God, what good would it really do? Perhaps a few well-trained people could understand it but, for most individuals, it would only have the force of someone else's assertion that it was legitimate. The majority of people would accept or reject the proof based on how much confidence they placed in the reliability of those who proclaimed themselves qualified to interpret it. At the end of the day, such a proof would probably have little meaning for most people, even if they believed it to be true, simply because it would be too far removed from things they do understand.

But who is supposed to be able to understand God? Should it only be the theologian after a life devoted to diligent study? Should it only be the quantum or astrophysicist who supposedly understands how nature (and, therefore, presumably God) works at the smallest or largest scales? What, then, about average persons? Are they just out of luck? What about the individual unable to get an education? (And how much education is enough?) What about a child?

The important thing about discovery is that it can proceed at many levels and has significance for the participants. We may be interested in the insights and experiences of others and may even understand them fairly well. Ideally, they will inspire us to engage in our own bit of exploration, for the discoveries that will have the most meaning for us are those we make ourselves. Even if on some grand absolute scale our insights don't measure up to those of the seminal religious figures, they are, nevertheless, the ones that will have primary significance in our lives. Some religions, though, go so far as to claim that the most important discoveries—that is, those at the highest levels of our imagined scale—actually are potentially understandable by everyone. This may not endear those religions to people who would like to think that their mental prowess should somehow qualify them for God's company[36] but it fits with the idea of an impartial God who only wants a relationship with those who want it with him and it suggests that being an explorer is crucial to that end. None of this means that the process of discovery should be short-lived—that once we obtain the most important insights we are done. As I've tried to stress, there should be much of interest to learn about a God with infinite attributes although the "most important" may be to acknowledge this, simply because it is the gateway to all the rest.

---

36. For the apostle Paul, we are all equally dim-witted before God: "the foolishness of God is wiser than human wisdom, and the weakness of God is stronger than human strength" (1 Cor 1:25).

CHAPTER 12

Is This Really What Religious People Mean By Faith?

> When faith is described as an element in culture and history, its nature tends to be grossly oversimplified, despite the vast and unconsulted literature of religious thought and testimony.[1]
>
> —MARILYNNE ROBINSON

BY NOW YOU MAY be thinking to yourself, "Well, all this makes sense but that is not what religious faith means to me. Faith means trusting even though I don't understand God or know how God works or can even imagine how something is going to work out."[2] Well, sure, as long as what you are talking about is not merely wishful thinking, there is no reason not to call that faith. But, if it is faith, it will still have all of the characteristics that we have discussed so far and deserves the same considerations. The critical aspect of faith such as this is that the probabilities associated with believing that God will act in one's best interests (even in the absence of any direct confirmation) are (presumably) based on faith that God has done so in the past. For this faith to be reasonable, sooner or later one must come to a trail of evidence.

---

1. Robinson, *Absence of Mind*, 14.
2. This is reminiscent of Rader's position described in an earlier chapter (Cory, *Quotable Quotations*, 129).

Faith Substantiated

This seems to be just the kind of mental state which characterized a significant subset of the Israelites of Old Testament times who, no matter how dire the straits (beleaguered by enemies, carried into captivity, etc.), nevertheless felt that God would somehow, at some time, rescue them and restore them to their former place of favor—an attitude based on their recollection of what they believed he had done during the exodus from Egypt many years earlier. Christians see the resurrection of Jesus in the same light—that is, as a manifestation of God's activity and a promise of what he is capable of doing. For the Christian, a God who can raise someone from the dead can do most anything. When a resurrection account is taken as evidence, particularly when supported by additional evidence, it makes perfect sense in light of that evidence to trust in the power that made it possible.[3] Most theists of any persuasion employ similar reasoning for any deity who can create a universe from nothing. A minor (or major) miracle here or there is relatively petty compared to such a feat. Such faith makes little sense only if there is no creator to begin with.

Note that this is far different from thinking that faith for faith's sake is somehow commendable. Faith is always in or about something—there is no such thing as raw faith. In religion, as elsewhere, the objects of one's faith and the significance of those objects, coupled with the depth and accuracy of that faith, are the arbiters of its merits.[4] Consequently, no matter which event or, better yet, series of events they rely upon as evidence, theists see their trust in God as being grounded in historic precedent—that is, as faith substantiated. Perhaps this is what the apostle Paul had in mind when he referred to "a righteousness that is by faith from first to last,"[5] or what Jesus meant when he told his disciples to, "at least believe on the evidence of the works themselves,"[6] or what the writer to the Hebrews was thinking when he defined faith as, "the substance of things hoped for, the evidence of things not seen."[7]

---

3. This is similar to Lewis's idea on authority (*Mere Christianity*, 53).

4. For example, when Jesus asked if he would "find faith on the earth" (Luke 18:8) the question was about the content of the faith he might discover and not whether there would or would not be faith—the existence of faith (in something) is a given.

5. Rom 1:17 (alternatively translated "from faith to faith" [King James Version]).

6. John 14:11.

7. Heb 11:1 (King James Version). In other translations words like "assurance," "conviction," "confidence," "sure," and "certain" are used but in this case the King James phraseology is particularly revealing. "Substance" and "evidence" are the very things into which we would all (religious or not) hope to be able to sink our teeth. Commenting on this verse, David Berlinski makes the following observation: "We can make no sense either of daily life *or* the physical sciences in terms of things that are *seen*. The

Of course this idea of an edifice of faith—of faith in one thing receiving its support from faith in another—is not unique to religious faith and can easily be seen in all applications of faith, from science to relationships. Because faith has a predictive quality, this often leads theists to formulate ideas about what they believe God will (or should) do based on what they believe he has done. This predictive character is probably closely connected to what many people mean when they claim that their religious faith is somehow different from other faith. Because all faith has a predictive nature, however, it is the connection to the object of the faith (i.e., deity) that makes it look different.

Clearly the predictive nature of faith is crucial to any religious undertaking (as it is to all endeavors) but, when the predictions go wrong, they make the whole enterprise look suspect. Frequently this is taken to imply that the entire foundation of that faith is ill-advised even though an equally plausible explanation is that almost any attempts on the part of individuals with limited understanding to predict in detail what a omniscient, omnipotent deity will do are bound from the outset to be problematic (and the more detailed the predictions, the greater the potential for error). Obviously one could construct predictions so general as to be vacuous—horoscopes come to mind—but that need not imply that all predictions are doomed. Most religions will contain predictions that are accepted by the majority of its adherents although the most commonly believed are likely to be those that refer to general occurrences or states of affairs rather than to specifics (e.g., God's ultimate triumph versus its methodology or a meaningful afterlife versus its particulars).[8]

This is often taken as a criticism of religious faith and sometimes that criticism is probably justified. Faith that deity has and will behave in a certain way is all about tracing a trajectory through religious space and there is always the danger that some people will plot the trajectory they crave, then try to make God fit it. With too general a trajectory, any deity will do but it is wrong to thereby equate general with spurious. Theistic faith with any merit, for example, must be specific enough to rule out the possibility of no God and will have definitive things to say about the attributes and behaviors of God. At the same time, most theists are probably careful to acknowledge that there is a difference in believing that God *can* do something and in

---

past has gone to the place where the past goes; the future has not arrived. We remember the one; we count on the other. If this is not faith, what, then, is it?" (*Devil's Delusion*, 45, italics in original)

8. Jesus, for example, while referring repeatedly to the prospects of final judgment, gave meager details about end times. On the other hand, competing interpretations of the book of Revelation (cf. Robbins, *Revelation of Jesus Christ* and Lindsey, *Late Great Planet Earth*), provide a classic example of an inability to reach a predictive consensus.

believing that he *will* do it. Be that as it may, one might expect that any viable religion would contain meaningful predictions that have had occasion to be verified. In assessing those predictions it is important to check that they were not added after the fact and also to realize that fabricating a prediction after something has transpired is far different from later recognizing that there was evidence all along pointing to the occurrence of that event.[9] One should also recognize the possibility that a prediction might be true merely due to chance and, in doing so, evaluate the relative significance of isolated predictions versus those that occur in the context of a broader story (e.g., as relayed in the scriptures of some religion).

On the Nature of Childlike Faith

Religious belief, like any other, consists of faith in events that are taken as evidence in support of further beliefs about the issues under consideration. However, there are additional reasons some persons will have for objecting to my lumping religious belief with other manifestations of faith. The picture I have painted thus far, in which I have promoted an intimate relationship between faith and reason and have extolled the virtues of active, exploratory search that can help increase the probabilistic accuracy of one's faith, may seem out of reach for many. Is the faith about which I have been speaking only possible for well-educated, highly motivated individuals? What about those religions where some manifestation of faith is deemed not only achievable but mandatory for everyone? I touched on this issue previously when discussing how the process of discovery can be undertaken at a variety of levels and yet remain meaningful and productive, but there is more.

When Jesus said that, "anyone who will not receive the kingdom of God like a little child will never enter it,"[10] and Paul noted that few of his converts, "were wise by human standards,"[11] they seemed to make it clear that, from their perspective, a childlike faith was both necessary and sufficient to win God's favor. That, I imagine, is what typifies many views of religious faith and there is no reason to quarrel with those views unless, by "childlike," one means uninformed, naïve, or (especially) blind. As a child, I trusted my parents precisely because they had proven themselves reliable (and not because I was uninformed, naïve, or blind). My children trust me

---

9. In science, for example, the initial evidence used to support Einstein's general theory of relativity was available for centuries prior to formulation of the theory but no one would have known what it meant, much less thought to look for it, prior to that.

10. Mark 10:15; Luke 18:17; cf. Matt 19:14.

11. 1 Cor 1:26.

for the same reason. Even a child's faith is based on evidence. Abused children develop faith that they will be mistreated in the same way that children with loving parents learn the opposite. Just because they are held by children, such beliefs need not be either wrong or feeble.

Trust in a parent is a relational faith and this may be the aspect of a child's faith that Jesus and Paul commend. Clearly this view is based on the belief that God has personal attributes but, if one can clear the anthropomorphic hurdles, their position makes rather good sense because knowing about God and knowing God are not the same thing. What if, before committing to marriage, I had said to my not-yet fiancée, "Dear, I would like to marry you, but first I feel compelled to understand how you work. How do your neurons fire at just the right time to generate feelings of love and concern? What is the interrelation between electrical impulses from your brain and the various ways your body functions? What genetic configuration has led to your green eyes and brown hair and sweet disposition? How are you able to distinguish between desirable and undesirable objects and in what ways have those distinctions led you to want to marry me (despite this pedantic list of questions)?" It probably goes without saying that one reason we are married is that, intrigued as I am by such questions, I managed to refrain from asking them (much less making them the focus of our relationship). Nevertheless, despite the ludicrousness of such a scenario, there are individuals who think that one must first understand what makes God tick before committing to (or in lieu of) a relationship with him. This is a rather curious state of affairs when—with apologies to my wife—God is a bit more complicated and harder to fathom than she is.

Emotions and Faith

Relationships, we know, are accompanied by feelings and it is that which animates many people when they refer to the specialness of religious faith.[12] The feeling component of faith is easy enough to understand. The bizarre imagined conversation in the preceding paragraph makes no sense, yet everyone understands the faith someone has that a specific marriage is in his best interest when the proposal emphasizes the fiancé's sensitivity, empathy, and sensuality. "Dear, I want to marry you. You make me *feel* so alive!" But where do the feelings come from? Those "good vibrations" that the Beach

---

12. William James (*Varieties*) believed that feelings were at the heart of religious experience and that logic takes a back-seat.

Boys sang about[13]—what we call "chemistry"—is impossible without contact, interaction, and a history of experience.

Certainly it is easy to protest that the evidence in this case comes from a genuine flesh and blood individual whereas, with deity, one is faced with determining whether any perceived interaction is with the real thing or just an idea, for even ideas can generate emotional responses.[14] The trick is defining "real thing" because, as we've seen, we cannot (or at least should not) expect God to necessarily affect us in the same way nor to provide the same kind of evidence that a person does. The mistake that many religious people make is to believe that the feelings accompanying religious faith somehow transcend rationality—a belief paralleled by the nonreligious who think that religious faith is impervious to rationality.

High IQ record holder Marilyn Vos Savant thinks that "matters of the heart" are different from "matters of the head" and thus make it "possible to hold conflicting beliefs."[15] Presumably, she knows that it is really all in the head but, even if her claim is true and we can hold conflicting beliefs under any circumstances, is it something we should permit to happen? I, for one, don't want separate areas of my brain believing two different things that are in conflict.[16] I don't find that idea—a sort of polygamy of the thoughts—acceptable in the least.[17] Nevertheless, William James thought that, "feelings of reality . . . are as convincing to those who have them as any direct sensible experiences can be, and they are, as a rule, much more convincing than results established by mere logic ever are." Reason, he goes on to say, "will fail to convince or convert you . . . if your dumb intuitions are opposed to its conclusions."[18]

But does James's judgment apply universally, or is it only mostly true or in need of some revision? C. S. Lewis, who describes faith as, "the art of holding onto things your reason has once accepted, in spite of your changing moods,"[19] seems to suggest that rationality can gain the upper hand. A little thought should provide plenty of examples that this is, indeed, possible. Virtually everyone has intimate experience with deferred gratification where gut

---

13. Wilson and Love, *Good Vibrations*.

14. As any casualty of unrequited love knows, sometimes it is only the idea that drives human relations as well.

15. Vos Savant, "Ask Marilyn," (2000) 22.

16. Cf. McGilchrist, *Master and His Emissary*.

17. Supposed incompatibilities between scientific and religious perspectives (cf. Krauss, "God and Science Don't Mix") are an example. A proper understanding of faith in general (and not just religious faith) is critical to any attempt at harmony.

18. James, *Varieties*, 46.

19. Lewis, *Mere Christianity*, 119.

instincts and cravings are subdued in favor of something rationally believed to offer a greater good.[20] Reason over feelings empowers every diet, exercise regime, sacrifice of sleep to study or work or care for children, willful abstention from sex under the wrong circumstances, and so forth. Reasoning takes into account how one will feel afterwards but the process is a rational one. Subjugating emotional drives to reason or employing reason to suppress one set of feelings for another may be difficult, but there is no mandate that the significant exercise of rationality should be easy.

Furthermore, the notorious unreliability of feelings should make anyone skeptical of putting too much faith in the message they purportedly convey. Speared by the nose of a fellow kayaker's boat several years ago, I felt sure my ribs were broken (they weren't). After sustaining a fall in Colorado's Black Canyon I was confident that my leg was not broken (it was). Most of us have plenty of stories where feelings have failed us or biased our tendencies to believe various things. Our feelings and the beliefs they engender thus span the cognitive depths where in the shallows something as basic as temperature judgments are affected by fluctuating metabolism while at the deep end feelings of worth can rise and fall with changes in hormones.[21] Music is especially efficacious at taking us on flights of fancy where it is easy to believe that all is right with the world whereas certain drugs have become famous for making people believe they could literally fly. Feelings need not be wrong but there is no reason to believe religious feelings are any more reliable than others—as the prophet had it, "Woe to you who are complacent in Zion, and to you who *feel* secure on Mount Samaria."[22]

Ultimately, when it comes to faith, knowledge trumps feelings. The best way to have feelings that are reliable is to have knowledge that is reliable. For faith to rest on a sound foundation it is not enough to feel that it does—it requires the knowledge that comes with evidence. There is no need to ask a life-long resident of New York if she feels like a citizen, homeowner, parent, accountant, etc. Either she is or she is not any of those and it is not likely

---

20. Perhaps this is partly what Polanyi had in mind when he said that, "The freedom of the subjective person to do as he pleases is overruled by the freedom of the responsible person to do as he must" (*Personal Knowledge*, 309).

21. The Christian Lewis noted that, "unless you teach your moods 'where they get off,' you can never be either a sound Christian or even a sound atheist, but just a creature dithering to and fro, with its beliefs really dependent on the weather and the state of its digestion. Consequently, one must train the habit of Faith . . . The first step is to recognize the fact that your moods change" (*Mere Christianity*, 119). Clearly this applies to the feelings accompanying all religious beliefs and not just those of Christians who, however, rather than being instructed to describe feelings are told to have a *rationale* for their hope (1 Pet 3:15).

22. Amos 6:1a, italics mine.

she spends much time agonizing over her feelings about them. The greater our confidence in something the less significance we attach to feelings about it.[23] Confidence, of course, is built on evidence. It is relatively easy to see this when it comes to love. One may start with the heart (i.e., fall in love) but unless there is evidence that the person is worth loving and that the love is reciprocated it cannot last. This may be one reason Jesus connected love of God by the mind with love from the heart, soul, and strength.[24] Only those who love God with their minds will have the proper long-term basis for loving him with their hearts.[25]

It is easy enough to downplay the role of rationality in matters of religious faith just because there has been, is, and, no doubt, always will be some poster child for irrational religious belief—a person obviously driven more by emotion that reason. But to assume that this automatically undermines the potential for reason to rule is to be guilty of the inductive fallacy.[26] Nevertheless, the tragedy is that all too often, religious or not, we do permit our reason with respect to matters of religious faith to be held hostage by our feelings. It is problematic, however, to make those feelings the defining characteristic of religious faith unless that faith is only capable of effecting the equivalent of a drug-induced ecstasy. Making emotion the main characteristic of religious faith, then, detracts from its ultimate significance unless its ultimate significance is merely to make one feel good.

None of this is meant to downplay the importance of emotions but to put them in their place (particularly with respect to religion). Finding their place, however, is not as straightforward as it may seem. Psychological and neurological evidence suggest a tight connection between emotion and reason[27] and most of us can appreciate how there can be a certain logic, even in feelings. For example, anticipating the feelings one assumes will be generated by engaging in some activity can become the motivation for doing so. Thus you, considering how you would feel after being caught, might believe that it is illogical to steal, but a person who is inclined to do so may think

---

23. Ravi Zacharias says it this way: "As godlike in their origin as feelings are, we must also learn to put them in perspective and protect ourselves from the glorification of feelings as the final affirmation of truth." He goes on to note that, "feelings could be the most seductive force to take us away from the truth" (*Cries of the Heart*, 45–46).

24. Mark 12:30.

25. This will sound backwards to many but the operative phrase is "long-term basis." One can fall in love with God during an emotional high (as with anything else) but only if one thinks God is worth loving will that person continue do so with the heart. As noted, feelings are valid when there is justification for them.

26. Hopefully, it is apparent that feelings are associated with rational beliefs just as much as irrational.

27. Cf. Damasio, *Descartes' Error*.

that the thing to be stolen (or the experience or challenge itself) will provide feelings that make it worth the risk.

On the Growth of Religious Faith

If it was a childlike faith that Jesus and Paul were commending, it is difficult to think that they meant it to be primarily about feelings. There is also evidence that they did not intend for individuals to remain stuck at some infantile level of religious faith. Childlike faith may be the starting point—what other kind could there be?—but it doesn't have to be where one stops. Children, after all, are expected to grow. This seems to be clear from Jesus' injunction to ask, seek, and knock.[28] As we've seen before, it is hard to miss the exploratory imperative in these words and these are activities that a child's natural inquisitiveness typically makes easy.

A child so engaged is almost sure to progress but it is not guaranteed. A lack of progression may simply signify a lack of engagement. Paul's comments to a congregation about giving them, "milk, not solid food" because they "were not yet ready for it,"[29] sounds like the technique any teacher would take when introducing a new concept but he is forced to conclude, "Indeed, you are still not ready."[30] The more sophisticated message he would have liked to share[31] was not within their current grasp. Paul summed up his desires for their growth by recounting his own experience: "When I was a child, I talked like a child, I thought like a child, I reasoned like a child. When I became a man, I put the ways of childhood behind me."[32]

When it comes to building a more mature faith, the rationality involved may have as much to do with seeing and interpreting the results of some starting nugget of belief as it does with a sophisticated chain of reasoning. When his disciples wanted him to increase their faith, Jesus responded by telling them, "If you have faith as small as a mustard seed, you can say to this mulberry tree, 'Be uprooted and planted in the sea,' and it will obey you."[33] Having failed to observe many displaced mulberry trees, I am inclined to think that either Jesus was mistaken or that he meant something beyond a literal way to make landscaping easier. Keeping in mind that Jesus

28. Matt 7:7.
29. 1 Cor 3:2
30. Ibid.
31. 1 Cor 2:6–10.
32. 1 Cor 13:11.
33. Luke 17:6. Elsewhere, he tells them that equivalent faith could move a mountain (Matt 17:20).

was speaking to people who had already seen sufficient evidence to support their faith (or they wouldn't have asked), is it conceivable to think he is telling them that, in some sense, they had enough faith already—perhaps in the sense of having the requisite starting amount? Was he saying that, by using the faith they had, they would experience results that would enable them to believe more (or that would raise the probabilities of what they did believe or would open doors to undreamt possibilities)?

I think it is important to recognize that if the seed of faith with which one starts is ill-founded, exercising that faith can cause it to decrease, rather than increase.[34] For anyone interested in truth, this should be a welcome conclusion but observation suggests that not everyone embraces such a thought. It is also the case that a request for an increase in faith could be nothing more than a desire for greater assurance in what one already believes without concern for its truth, or simply a guarantee that what one wishes to be true really is.[35]

There is, however, something of a paradox here in that recognition of the limitless possibilities made available by faith in God must be done at the expense of acknowledging personal limitations—in a word, of knowing one's place. When growing up, few harbor delusions about the respective roles in their families of parent and child. Jesus' aforementioned approbation of a childlike approach to God reminded his disciples that, no matter how impressive they might look to each other, their credentials were rather paltry when compared to an infinitely capable God. When people see faith in God as somehow different from their faith in other things, it may be that this is one of the things they have in mind.[36]

Rethinking Pascal's Wager

Ultimately, then, what makes religious faith—specifically theistic faith—appear to have a different character is not the faith itself but the object of that faith. Even the strictest of atheists is likely to admit that, if there was a God, he might possess attributes so far beyond our own as to render faith in him something special. That, after all, is precisely what they are arguing against.

---

34. Of course, for every decrease of faith in item "X" there is an accompanying increase of faith in "not X."

35. It should be clear by now that little, if any, of this is limited to religious faith.

36. Childlike faith is only a partially informed faith but so is adult faith. Those previously referenced Bible passages (i.e., Mark 10:15; Luke 18:17; cf. Matt 19:14) are not the only places where Jesus is recorded as targeting a presumptuous arrogance on the part of his disciples and uses a child to make his point (cf. the synoptic parallels in Matt 18:3, Mark 9:37, and Luke 9:48; also compare 1 Cor 1:25).

When Jesus told someone, "Your faith has saved you,"[37] he clearly didn't mean faith by itself. It was not some free-floating, vague, detached belief but a very specific, directed faith (made visible by that person's actions) that he applauded.

It is this faith in deity that is at the heart of Pascal's wager. I introduced this earlier but let me restate it here:

> You must wager. It is not optional . . . Let us weigh the gain and the loss in wagering that God is.[38]

Those feelings of otherness that seem to accompany faith in God arise precisely because that faith is in something which, if true, is potentially far more important than the other objects of our faith—and the stakes are dramatically higher.[39] This is the thrust of Pascal's wager.

A number of years ago Marilyn Vos Savant received a letter from one of her readers asking what she thought of the case Pascal had made. Her three-pronged reply[40] denied the force of the argument, beginning with the apparent irrationality of Pascal's thesis in light of his presumed belief in the lack of free will. If that is an accurate characterization, it is indeed difficult to see much merit in his suggestion but there is reason to doubt that Pascal was as deterministic as Vos Savant claims.[41]

In any case, Vos Savant goes on to provide two other reasons for thinking Pascal mistaken. For the first of these, she paraphrases Pascal's thesis this way: "if God exists, believing in Him can bring eternal life," and then claims, "Pascal's argument implies that the best bet is to join whatever religion makes the most promises, because one has the most to gain if it turns out to be the correct one."[42] Although there is no guarantee that any god would necessarily feel a compulsion to reward people who merely bet on his existence (regardless of what Pascal or Vos Savant might have thought), that is not what I take to be the primary significance of Pascal's wager.

---

37. Luke 7:50.

38. Pascal, *Pensées*, #233.

39. By defining faith as "ultimate concern," theologian Paul Tillich (*Dynamics of Faith*) jumps directly to the significance of high stakes ventures. Unfortunately, his definition loses contact with faith in many of its other manifestations as described in this book.

40. Vos Savant, "Ask Marilyn," (1996) 20.

41. For example, Pascal's response to an imagined complaint from someone who says, "I . . . am so made that I cannot believe" (Pascal, *Pensées*, #233) does not sound like the words of a man bound by the chains of a strict determinism.

42. Vos Savant, "Ask Marilyn," (1996) 20.

Vos Savant is mostly correct if there is no way to verify religious claims. If the evidence for all religions is equally vague or equally sound, then—assuming equivalent requirements—one religion is as good a bet as another and, under those circumstances, one would certainly want to bet on the one with the biggest payoff. Even though Pascal framed his argument just in terms of belief or disbelief in God (and not in terms of multiple versions of reward), in this regard Vos Savant could be right. But, if we stop there, we stop too soon. What about evidence? With respect to God, Pascal himself claimed (in preface to his proposed wager) that humans are, "incapable of knowing either what He is or if He is."[43] This is a strange statement in light of his suggestion to look to Scripture as a "means of seeing the faces of the cards"[44] and gives the impression that Pascal was conflicted over the issue of evidence.

Perhaps that was the case but one also must allow that Pascal was merely providing a starting point for those who felt that it was impossible to know about God.[45] In either case, the wager is significant because it suggests the potential for a belief in God's existence to serve as the foundation for something more—to act as a kind of mustard seed, as it were.[46] As the writer to the Hebrews put it, "Without faith it is impossible to please God, because anyone who comes to him must believe that he exists and that he rewards those who earnestly seek him."[47] If that is right, belief in God's existence is nothing less than a gateway to exploration.

Michael Shermer has responded to Pascal's wager by saying that, "If forced to bet on whether there is a God or not, I bet that there is not, and I live my life accordingly."[48] As we've already noted, such responses to a perceived binary dilemma merely reflect some underlying belief probability (which is, most likely, in a state of perpetual flux for most of us) but those

---

43. Pascal, *Pensées*, #233.

44. Ibid.

45. Ibid. Pascal asks, "Who then will blame Christians for not being able to give a reason for their belief . . . ?" Well, in light of all that we've considered so far, I would, for starters. But so, I think, would the Christian author who admonished his readers to, "Always be prepared to give an answer to everyone who asks you to give the reason for the hope that you have" (1 Pet 3:15). One need not be a Christian to see the merit in that. If Pascal actually believed in our inability to know about God (rather than using his statements as straw men), then there is much for which to fault him. But, unless he believed the evidence to be better, why did he select Christianity over the alternatives? And if he selected Christianity (with its very specific claims about the character of God), how could he really believe that God "is infinitely incomprehensible?" Pascal, *Pensées*, #233.

46. Cf. Luke 17:5–6.

47. Heb 11:6.

48. Shermer, *Science of Good & Evil*, 4.

IS THIS REALLY WHAT RELIGIOUS PEOPLE MEAN BY FAITH? 235

responses are more than just our modus operandi, for they can open or close the door to subsequent insights and possibilities that might otherwise be available. In a very practical way, atheists and theists are both right up to a point—there either isn't or is a God as far as they are concerned. In a very real way (but not necessarily the most real), God only exists for those who believe that he does. For Shermer, even if he does, he doesn't; for the prophet Isaiah or the apostle Paul, even if he did not, he did. If there is no God, Isaiah, Paul, and quite a few others will have imagined a great deal about nothing; if there is, Shermer and others will have made it difficult for themselves to know much else about him. We may make our own reality but that doesn't mean it is real.

In their attempt to define virtue, an otherwise sagacious Socrates bullies Meno into agreeing that one cannot know a part of virtue without knowing all of it.[49] But that makes little sense when seen in light of our other experiences. For example, we can surely know a part of the world without knowing all of it. In fact, we can never know it all but any attempts to do so must necessarily consist of knowing parts (in succession) and the process results in our knowing more and more about the world itself. It is reasonable to consider the possibility that a similar insight applies with respect to knowing God. Biologist Harold Morowitz, writing about the "epistemic circle" that confronts the scientist, remarks that, "One starts with the mind as the primitive and goes around the circle of constructs in an effort to explain mind."[50] While it may have its flaws, Pascal's wager suggests that those who would know God start with a belief in his existence. If an "epistemic circle" seems likely to result, it cannot be claimed that it is something unique to religion.

Another thing we can note about Vos Savant's response pertaining to promises is that it focuses on the quantity of the presumed benefits rather than their quality. Clearly, however, it is not the number of promises or things promised but their value that matters. Given the choice of fifty canoes or one yacht, I'll wager you'll take the yacht. Naturally, any chance of your actually getting it depends on who is making the offer.

The issue of value is also integral to the final rationale offered by Vos Savant for rejecting Pascal's wager, which hinges on his, "assumption . . . that nothing is lost by a mistaken belief" and her assumption that, if he was wrong, he had wasted the latter part of his life because of his religious beliefs.[51] Perhaps, but Pascal himself noted that the person who took his

---

49. Plato, *Meno*, 44.
50. Morowitz, *Emergence of Everything*, 8.
51. Vos Savant, "Ask Marilyn," (1996) 20.

advice could expect to reap significant benefits during this life, *even* if he lost the long-term bet.[52] Regardless of who was right, both positions reflect beliefs about what is of value, something that has always been central to the claims of religious faith.[53]

Earlier we had occasion to think about Mencken's belief that, "The most costly of all follies is to believe passionately in the palpably not true."[54] In the first place, I doubt that there are many people who do that—undoubtedly a great many believe in things you or I think are blatantly untrue, but not that they themselves do. Without quibbling over that, however, we can now see that Pascal's wager, despite its possible weaknesses, gives us another way to think about Mencken's claim. Could it not be equally (if not more) costly to disbelieve in the true? This is the assessment of value that, I believe, Pascal intended when he proposed his famous wager and which is integrally connected to perceptions that there is something *potentially* exceptional about religious faith.[55]

---

52. Pascal, *Pensées*, #233.

53. Would it make better sense to believe in God if one thought it was not valuable to do so?

54. Mencken, *Mencken Chrestomathy*, 616.

55. For Biblical perspectives on value, compare Matt 13:44–46; Luke 16:15; and 1 Tim 4:8.

PART IV

Conclusion

CHAPTER 13

No End in Sight

We are not afraid to follow truth wherever it may lead, nor to tolerate any error so long as reason is left free to combat it.[1]

—THOMAS JEFFERSON

Factland and the End of Faith

A FEW YEARS AGO, Sam Harris produced a book entitled *The End of Faith* as an attempt to express his beliefs about (and disdain for) religious faith.[2] It's a catchy title but, although there may be an end of faith in specific things or perhaps even an end of particular kinds of faith, there can never be an end of faith unless there is also an end of us. I hope that, by now, this has become clear. Nevertheless, let's attempt to imagine a world where Harris gets his wish and faith is not only not required but simply does not exist—not just religious faith but faith itself. This will require a bit of effort because, as anyone who tries this will quickly discover, such a world is so far removed from the one which we actually know that it is all but impossible even to imagine.

What would it be like in a world where there is no faith? In short, everything would truly be a fact. In Factland the idea of probabilities would never occur to us because the result of every decision and the outcome of every anticipated action would be known before being made. The teenage boy would know the response of every girl he dreamed about asking for a date prior to

---

1. Jefferson, "Letter to William Roscoe" in reference to Jefferson's vision for the incipient University of Virginia.
2. Harris, *End of Faith*.

ever making a call. He might, therefore, elect not to ask in some cases but he would also know whether, if someone said "No," persistence would ever pay off. The teenage girl would know that any particular boy would or would not ask. In each case there could be no doubt. Furthermore, there might be multiple reasons for dating someone over a period of time but trying to determine if there was long-term potential in the relationship would not be one of those reasons—that would be known before the first date.

Similarly, there would be no need for polls because everyone would automatically know the outcome of an election before voting. Actually, voting itself would be superfluous. Simply knowing about an election and who the candidates are would be enough to know the outcome, so why vote? Of course, that means that potential candidates would also know the outcomes, so it is unlikely that anyone but the assured winner would even enter the race. That, obviously, makes knowing the outcomes easy for everyone else, not that they would need any help. If this strikes you as a bit confusing, welcome to the paradoxical world of certainty. In that world we would know nothing of bookies or gambling, insurance agents, or stock markets—they simply couldn't exist. In that world, every diagnosis is assured and the results of any possible cure known in advance. Only those treatments guaranteed to be successful would be selected, assuming there were any. Everyone would share the same conception of deity (although it would still make no difference to some) and no scientist would wonder about the fruitfulness of a potential experiment or financier about the wisdom of an investment. "Contingency" and "speculation" never enter the vocabulary and qualifiers like "perhaps" and "maybe" are found only in storybooks. Words such as "belief" and "trust" would be meaningless and hence unknown while terms like "guarantee" and "assurance" would merely seem redundant. Promises would be hollow and therefore never made. Surely there would be no books on faith.

Because chance can never enter the picture in such a world, the concept of risk is not only unknown, it is unimaginable. There would never be concern as to whether one had enough data to make an informed decision—any amount would be enough. One must merely be competent to ask the right questions, in which case the answers would then be known at once. This does not imply omniscience, for even in a world of pure fact we could only inquire into things of which we had some prior knowledge. We must still know what questions to ask. Paradoxically, this fact would likely go unnoticed by most of the inhabitants of our belief-free world, as the abundance of what they know for sure clouds their vision of what they do not know. Blindness to one's own ignorance is no different in that world than in this.

It is true that the questions posed must be those that could be answered yes or no, true or false. But every question can be changed (with varying degrees of success) into one or more questions of the "yes or no" kind and for any such question we could conceive we would know the answer. Want to know where to go on your next vacation for maximum enjoyment? Perhaps the answer is a small town in Idaho but, if you had never heard of it, you could not know by asking directly. However, you could ask a series of geographically limiting questions—Should I vacation in the Western hemisphere? Should it be in North America?—that eventually pinpointed your destination as closely as you liked. You would be your own personal oracle.

In the world we currently occupy there is a sense in which we often act as though the result is known—the teenage boy asks for a date believing his proposal will be accepted (and that belief along with a healthy measure of hope helps motivate his call)—but in the world I have been describing there is no angst over the decision to call or not, no fear of embarrassment that he will be rebuffed, and no thrill of excitement over the possibility of a positive response. In that world, you would never ask or be asked "What do you think?" or "Can you imagine?" and with the end of faith would also come the end of hopes and dreams for they, too, would no longer be needed in such an immensely practical but utterly bland world.

When we are only faced with choices for which the result is known we can hardly call them choices. Technically, you could select the harmful or less appealing ones if you wished but that would be like hitting yourself in the head with a hammer—it is a choice but who would make it? With no intermediate probabilities for anything, one will have discovered a unique kind of high-level determinism far more insidious than any currently conceived for the universe we actually inhabit. If you're finding it difficult to imagine living in such a universe, that is because you can't imagine living without faith and you can't imagine that because there is no such place.

I have posed this mental exercise to demonstrate the absurdity of thinking that faith is something we can expect to go away or that we can discard as our species matures. Unless we are or become omniscient—which is more than even the assumption of a fact-only world warrants—we will always rely on faith to understand and interact with whatever world in which we find ourselves. Furthermore, it is and always will be through faith that we are actually enabled to find ourselves.

Moreover, it turns out that even in our imagined world of nothing but facts, faith must still exist. We can see this in several ways. In the first place, you will recall that, in the world we have been considering, the future remains unknown with respect to any particular issue until we ask the proper questions about it. What, however, are the proper questions? Presumably, *that* is a question that must precede the ones we actually intend to ask. The possibility of an infinite regress seems quite real in this case but the existence of even one level of what we might call "meta-questions" suggests that the future we anticipate coming to see must necessarily follow some hypothetical future we have already envisioned. But until we ask the right questions, that vision is incomplete and any bias toward one possible outcome over another reflects a probability we thought impossible in such a world. In other words, belief would always run ahead of knowledge, even in a supposed fact-only world.

That's not all. Even in that world of facts, we could never ask every question needed to clarify all of the issues that confront us. Presumably, for instance, knowing that we would come to harm on a trip would be sufficient to cause us to choose not to go—if, that is, we knew how to assess "harm." Suppose some hypothetical traveler knew that he would sustain a broken leg if he went on a climbing trip proposed by friends. That might be an adequate deterrent except further suppose that, if he went, he would meet his future spouse in an emergency room. Did he think to inquire into that possibility? Would that knowledge have been enough to change the verdict from "no go" to "go"? Would it matter if it was his neck instead of a leg that was broken or if she had one set of personal attributes instead of another?

There is no end to the potential for such questions. Consequently, our wary traveler may try to sidestep the dilemma we have posed for him by asking only the most general questions—for example, "Will I ultimately be happier by going on this trip than if I do not?" or, better yet, "Is this the best thing for me to do?" Now the second question is a good one for anyone to ask but, even if he gets an affirmative answer, there is no reason to think that he will know why it is best. What turns out to be in his best interest may not be best for the girl he meets, his traveling companions, or anyone else for that matter. Concepts such as "happiness," "best," "good," and "bad" are almost always relative to some particular set of circumstances, although the context of life is so nearly infinitely complicated (by individual actions as well as by all the other persons and situations encountered) that one cannot know the final significance of those concepts until the game has been played. By then, things (and views) will have changed. As a result, even the fact-obsessed traveler cannot evade the one clear fact that he will still harbor beliefs about the meaning of his visions.

In a final attempt to escape the grip of faith, some inhabitants of Factland may merely refuse to frame any questions about the future. However, they will then be fully open to an imagined one—that is, one about which they will find it difficult not to have beliefs. There is no escape.

Why, then, might one struggle to salvage such a world, even though it is a losing battle? Here we run up against a paradox in that we want certainty but not too much because we do not know how to reconcile omniscience (including and perhaps especially our own) with freedom. On the one hand we have a genuine desire to maximize the accuracy of our probabilistic assessment of things while, on the other, we have the sneaking suspicion that, if we were wholly successful, we would be little more than automatons. We wish our beliefs to be correct because that offers the best chance for successful and meaningful living but Factland, while initially attractive, seems eerily confining.[3]

Part of this conundrum is manifest in an age-old fascination with fortune-tellers and prophets,[4] a lure that loses its appeal when it is realized that knowledge of the future entails an unavoidable destiny. Is it really natural to seek restrictions on one's own freedom or is doing so an implicit acknowledgement that there truly is none? I imagine that this is seldom considered and that one of the main reasons for seeking clairvoyant insight is simply a matter of comfort. Perhaps this is why some of us watch re-runs of old football games in which our favorite team was dominant—with a prior guarantee of victory, worry over the outcome is impossible. It is reassuring to know that things will go our way but in this case the knowledge is really about the past, not the future, and freedom is gone.[5] Such a game can happen only one way and there is nothing anyone can do to change it. But any full certainty about a real future would be no different.[6]

This has been an intellectual dilemma at least since the time of Isaac Newton, whose convincing arguments for believing in a mechanistic universe seemed to spell the end of any true freedom.[7] The argument is simple—given sufficient computing capacity and complete knowledge about

---

3. Wendell Berry imagines another restriction of freedom: "But if you had complete knowledge, if you knew everything, could you then act? Could you apply what you knew, or would you be paralyzed by a surplus of considerations?" (*Life is a Miracle*, 149). In fact, he goes so far as to call it a "sin" for one to wish "that life might be, or might be made to be, predictable" (6).

4. Wizards, mediums, and crystal balls aside, the natural awareness of our limitations to predict the future is really just an implicit acknowledgment of the probabilistic character of our beliefs about it.

5. Along with excitement, anticipation, interest, and hope . . .

6. This extends to concerns about what is expected when praying for guidance regarding some pending decision. If given an oracle about every choice, would it lead to contentment or concerns that autonomy had somehow been sacrificed? Perhaps most such prayers are really requests to be pointed, not pushed.

7. So far, the randomness inherent in quantum theory has been of limited usefulness in resolving this issue.

the current state of the universe, we could predict everything. This position, however, articulated by Pierre Simon Laplace[8] many years ago and repeated innumerable times since, is an "in principle" observation that can never be true in practice, as modern scientists have made clear.[9] "In principle," any number of bizarre and disturbing things *could* happen but that doesn't mean there is much danger of them actually doing so and we should be wary of building our metaphysics around them.[10] We may have trouble deciding what it means for our freedom if God knows everything about the future but there is no need to worry that we can ever know that much.

We can, therefore, make knowing everything with the greatest possible certainty a goal without any fear that we are undermining our freedom by doing so, simply because it is a goal we can never achieve. Trying to achieve it, however, can actually increase our freedom in dramatic ways by enabling us to understand and accomplish things not otherwise possible.[11] This is the position in which we find ourselves, poised between an impossible dream and a debilitating ignorance.[12] Ours is a dream which, if reached, would seem more like a nightmare but which, when pursued, opens the doors to the most meaningful existence possible.

I know of no easy answers to the free will conundrum but would like to suggest that if true freedom exists it enters the picture when we learn that, in order to make the probabilities associated with our beliefs high, we must seek evidence and learn how to evaluate it (and then actually do so). Factland is a myth—we cannot escape faith nor should we want to. But, if we have any freedom at all, we can choose not to live for long with a faith that lacks substance or ignores evidence. Some development along these lines may happen more or less automatically but most will not without a conscious effort to make it happen. Consequently, such a conscious effort is both one of the prime manifestations of any free will that might exist and one of the reasons to think it does.[13] To look for evidence and evaluate it is

8. Laplace, *Probabilities*, 4.

9. Cf. Hawking, *Brief History of Time*, 167; Wolfram, *New Kind of Science*, 1135; Lloyd, *Programming the Universe* (e.g., 98).

10. As we've already seen, being wary doesn't mean that the possibility should be discounted off-hand.

11. The freeing power of restrictions has been noted with respect to such things as language (Pinker, *Language Instinct*, 292), music (Stravinsky, *Poetics of Music*, 65), human freedom in general (Peterson, *Minding God*, 97), and evangelism (1 Cor 9:19).

12. This precarious position may just be the one that provides the greatest likelihood for meaning and value. A provocative physical analog is Kauffman's (*At Home in the Universe*) "edge of chaos."

13. Cf. Ward, *Big Questions in Science and Religion*, 129: "Since physical determinism cannot be proved and since indeterminism is affirmed by most quantum physicists

one of the most (if not the most) important choices we make with respect to our potential for freedom.[14] If there is any such thing as real freedom, it will be found most noticeably in the process of exploration.

How Can Truth Seekers End Up in Different Places?

When I was a boy, I had visions of becoming an astronaut. The prospects of space travel fascinated me and continue to do so even now. But space is more than a synonym for the cosmos, and other types of space also deserve exploration. The difference between an "atheist" and "a theist," for example, is just a space. Ostensibly, atheists and theists are as different as night and day, with each accusing the other of being in the dark. But, are they really so far apart? What's in the space?

To answer these questions, it is important to consider that evaluating similarity and difference depends on how we choose to view things. Linguist George Lakoff has illustrated this nicely in his book *Women, Fire, and Dangerous Things*[15] where he discusses our ability to form categories composed of seemingly disparate items based on observations of commonalities that extend beyond surface features (Any guy who has been burned in a relationship will immediately understand his choice of title). When it comes to atheists and theists, we can hardly dispute the significant differences in the content of their beliefs—the objects of their faith and the significance (depending upon who is right) of those objects vary dramatically. And, even though they may each hold onto their faith with equal strength, the accuracy of their beliefs must necessarily differ. However, taking a hint from Lakoff, we might wonder if atheists and theists have more in common than meets the eye. In particular, we can ask how it is that an individual comes to be one and not the other. Is it possible that similar forces are at work in crafting each? Is there a sense in which believing or disbelieving in God is much the same? What's in the space?

Several hundred years ago, Francis Bacon noted that, "a little philosophy inclineth man's mind to atheism; but depth in philosophy bringeth

---

anyway, it looks as if a consideration of human responsibility gives good reason for affirming a processive view of time, in which open futures are at least partially fixed by the choices of rational agents within the process."

14. One might also postulate that the urge to look for evidence is simply a deterministically derived drive that happens to some people but not to others. Under such circumstances, however, any call to personal decision would be subject to the same critique.

15. Lakoff, *Women, Fire, and Dangerous Things*.

men's minds about to religion."[16] Many people would argue that the roles of "atheism" and "religion" in Bacon's statement should be reversed but, in fact, it happens both ways. Intelligent, educated people have proposed what they consider to be rational arguments for and against God, with each side suggesting that the other is ignorant and irrational.

Throughout this book I have promoted the idea that exploration is the key to developing and maintaining a coherent and meaningful set of beliefs but neither freedom to explore nor exploration itself guarantees success. If we are really free, we are free to make mistakes. Shortly after the well-known author and defender of the atheist faith, Christopher Hitchens, died, a Christian named Larry Taunton described their friendship.[17] In his essay, Taunton noted that, at the outset, "We immediately discovered that we had much in common," and went on to describe a road trip in which they had "a civilized, rational discussion" about the Gospel of John. As a man who "disliked unintelligent conversation," it is unlikely that Hitchens would have spent so much time with Taunton had he not felt that his friend was capable of intelligent communication. So, if both men truly had so much in common and really were intelligent, how could they end up with such different world views? How is it that an E. O. Wilson or a Michael Shermer can grow up expressing belief in God and end up eschewing it while crossing paths with a Francis Collins or an Antony Flew who started at the opposite pole? Figure 1A illustrates this scenario but it is equally easy to imagine other possibilities with respect to the trajectories that the beliefs of any two persons can take regarding the same subject. In Figure 1B, for instance, two individuals begin with similar beliefs but end up believing quite different things. Alternatively, the starting point for one of the parties may never change. Ideally, the beliefs of truth seekers would converge to a common conclusion no matter where they started (as in Figure 1C) yet, all too often, that seems not to be the case. But if a particular view concerning the existence of God is truly correct, if one form of government or economy actually is better than all others, if there are a collection of generally beneficent morals, if one scientific theory can be closer to the truth than another, then how can an honest, effortful search for those best answers take people to such different destinations? Can truth seekers really end up in different places? How can that be?[18]

16. Bacon, "Of Atheism," 40.

17. Taunton, "My Take."

18. Although I have asked these questions in the context of a religious example, they are relevant to any area of belief.

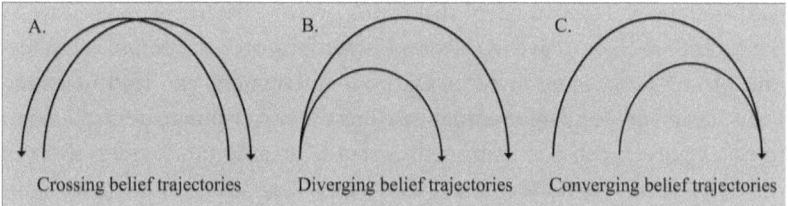

Figure 1. Possible paths for the beliefs of two individuals

Well, we can begin by acknowledging that the implicit assumption in these questions—that is, that everyone is really seeking truth—is probably unfounded.[19] As already noted, the inherent difficulty of a truth-seeking venture can deter such a search and, when one admits that some portion of a current belief could be wrong, the impending task may become too much to bear. Furthermore, the mere proclamation that one is seeking truth—following the evidence wherever it leads as Antony Flew was fond of saying[20]—is no guarantee that one is actually doing so. It may be nothing more than an attempt to appear sophisticated. Even if all truth seekers were somehow instantaneously color-coded for identification, we would still be unsure about how many there are because truth-seeking is not likely a monolithic property of any person. More probably, it comes and goes with a changing environment of influences and events and almost certainly with the topic at hand, leaving even the best truth seeker among us looking somewhat like a chameleon.

Perplexity about the potentially different destinations of truth seekers is also due to a tendency to believe that anyone seeking the truth must somehow be assured of success. Although this assumption is inherent in many educational endeavors, and even though a religion may make such a promise under special circumstances,[21] there is no general guarantee. It is probably safe, however, to say that those who seek truth will be more successful than those who don't—that is, they increase the probability of their success. Occasionally people may stumble onto truth even when they are not looking. As often as not, it will go unrecognized because the criteria and motivation for recognizing it have never been adequately established.

If we limit our focus to genuine truth seekers, assume they are honest, consistent, and thorough, and acknowledge that success is not guaranteed, it is still tempting to believe that they should reach the same conclusions, at

---

19. Consider, for example, Giberson's assessment regarding evolutionary controversy: "Truth is valued when it serves a purpose and not for its own sake" (*Saving Darwin*, 166) and "the goal of the protagonists is to *win*, not to discover the truth" (172).

20. Flew and Varghese, *There is a God*.

21. Cf. John 8:31–32.

least eventually. Unfortunately, that is not always what happens, even when the stakes are high. Traveling without directions over the complicated terrain of a difficult issue or receiving poor instructions can lead to wrong turns, dead ends, and an eventual weariness that conspire to persuade even those explorers who start with enthusiasm to abandon their quest short of the destination. As focus wanes, it becomes ever easier to settle for convenience over truth and, at that time, the search is effectively terminated. Others are marooned by fear or, more commonly, diversion—for every Odysseus there is a Circe; for every Lancelot a Guinevere. Even if we persist, time can run out and leave us stranded.

Furthermore, simply believing one is rational is not enough. Throughout the years some of my students have been disappointed to discover that their supposed rational exposition of a problem was not. On occasion, I've been forced to admit that my imagined rational critique of their supposed failure was not. An individual who believes that she has reached her conclusions on the basis of intellectual inquiry is likely to consider them valid simply because she has tried to be rational. Trying is admirable—it is what explorers do—but the real effort expended is easily exaggerated. If the energy required to decipher evidence exacts a price she is unwilling to actually pay, the result can be an irrational defense of her presumed rationality.[22]

It is also the case that, as we've seen, the likelihood of ending up at the same place is compounded by the simple fact that almost never will any two people start at a common origin. Even if they do, they will always be confronted with different environments along their journey; their routes, therefore, must necessarily be different. Fortunately, although this may induce delays, it need not make the attainment of the same goal impossible. Far more sinister is the case where the presumed explorers have different goals, for then consensus can almost never be reached. If I believe that there is no such thing as a tasty, fat free éclair and want to defend that position at all costs, I am not likely to be found anywhere looking for one and may even refuse to sample what you set before me. Such obstinacy is appalling, even when the only concerns are cholesterol and flavor and it doesn't much matter. Unfortunately, the same thing can happen when debaters about something truly important—perhaps God or the environment—have made

---

22. It is one thing to have access to evidence and another to process it correctly. For example, consider, Jesus' response to a request for a sign: "You know how to interpret the appearance of the sky, but you cannot interpret the signs of the times." (Matt 16:3b). One might have expected Jesus to applaud an ostensible search for evidence but his response indicates that his interrogators had (or would have) all the evidence they needed. The issue, apparently, was what they would do with it and it was that which distinguished their reactions rather than the evidence itself.

maintaining their current belief the primary goal rather than attempting to know the truth (e.g., about God or the environment). If we decide ahead of time that our destination is a particular belief, we can reach it and still be wrong. When the destination is truth about the belief itself, however, then being wrong simply means we are still on the journey.

None of this will (nor can) preclude our hoping that one belief or another is the correct one or prevent us from having some measure of faith that it is. Consequently, it does not have to mean that we take no position—as Pascal reminds us, in some arenas we must wager. To be authentic, though, truth seekers are obligated to be entirely clear about their goals. The distinction here can be subtle. Being uncertain about something inclines us to waffle but choices have to be made and lives have to be lived.[23] Sometimes, however, directions need to be changed. Prematurely staking a claim can tie one so securely to what would otherwise be seen as an altogether untenable position that the necessary movement becomes essentially impossible. No wonder we end up in different places.

While the homesteading analogy is compelling, it is not the whole story for it gives the impression that truth-seeking ventures are stymied solely by the closed mind. But even exploration can stagnate if one is merely re-investigating the same kind of terrain. The result is a sinister kind of immobility that is difficult to perceive because there is an undeniable sense of activity.

When he was a teenager, my younger brother, who had a summer job in a rural community a couple of miles away from our equally small town, decided to hop on a slow-moving freight train to expedite his return from work one day. Unfortunately for him, by the time the train reached our town it was moving far too fast for him to make a safe exit. When it had finally slowed sufficiently for him to get off, he was many miles away.

In similar fashion, it is easy to be carried along by a particular philosophy, support group, social network, lifestyle, and so forth to such an extent that the perceived trauma of disengagement traps us as surely and tightly as if we were going nowhere (and indeed we may be). Thus, barreling along pell-mell we remain unchanged, having little chance to accumulate any truth that might exist along the terrain through which we pass. Moreover, if we fail to realize that an imagined shortcut has actually taken us further from the truth, we are doomed to end up at a destination we never intended. Cognitive momentum[24] propels one in a fixed direction as surely

---

23. Cf. Shermer, *Science of Good & Evil*, 4. The author of James put it bluntly: "the one who doubts is like a wave of the sea, blown and tossed by the wind" (Jas 1:6b). Perhaps it is not doubt itself that is the problem but doubt that precludes action.

24. Cf. Donaldson, "Predictive Learning and Cognitive Momentum."

as a speeding freight train and is just as difficult to derail.[25] Much of the time, this is a good thing but it is sobering to consider that at any moment and with respect to any matter we could be rushing away from the truth rather than toward it. Some destinations are better never reached.

In the light of these considerations, the whole truth-seeking venture begins to look perilously fragile. However, it is only when we are constantly aware of the factors that put us on a particular track or work to keep us there that we have much chance of knowing when we should reevaluate our itinerary. That, of course, is also a matter of faith but the real danger is that disillusionment with our own search can lead us to determine that truth is relative—a truly scary proposition because it probably means that we have given up on the quest altogether. Yet, no matter how thorough or advanced our journey—even when our physical journey is nearing completion—we have barely begun. When it comes to mental and spiritual excursions, we never arrive. This is not to say there are no useful and meaningful way-stations. Nevertheless, the moment we think we have arrived the journey is over, although we will not be at the final destination for there is none.[26] There will always be more things to explore. Thus, if we ever stop, we will have stopped short.

---

25. My favorite quip illustrating the vicious and sad (but sometimes funny—at least to observers) self-supporting merry-go-round that can typify some belief systems is from a sermon-spoof: "We know this is right because that's what we've been preaching" (attributed to Dr. James O. Combs).

26. As Bak has it, "Nothing prevents further progress more than the belief that everything is already understood" (*How Nature Works*, 120). Though stated in reference to evolutionary theory and the sciences in general, this is true for anything.

References

Abbott, Edwin A. *Flatland: A Romance of Many Dimensions*. New York: Dover, 1992.
Adams, Scott. *God's Debris*. Kansas City, MO: Andrews McMeel, 2004.
Albus, James S. "Outline for a Theory of Intelligence." *IEEE Transactions on Systems, Man, and Cybernetics* 21 (1991) 473–509.
Anderson, Chris. "The End of Theory: The Data Deluge Makes the Scientific Method Obsolete." *Wired Magazine* 16.07 (2008) 108–9. http://archive.wired.com/science/discoveries/magazine/16-07/pb_theory.
Aquinas, Thomas. *Summa Theologica*. Translated by Fathers of the English Dominican Province. New York: Benziger Brothers, 1947.
Aristotle. *On the Parts of Animals*. Translated by William Ogle. Internet Classics Archive. http://classics.mit.edu/Aristotle/parts_animals.html.
———. *Physics*. Translated by R. Hardie and R. Gaye. Internet Classics Archive. http://classics.mit.edu/Aristotle/physics.html.
Bacon, Francis. *The Great Instauration*. In *The Works*, Volume VIII, translated by James Spedding, et al. Boston: Taggard and Thompson, 1863.
———. "Of Atheism." In *Bacon's Essays*. London: Macmillan, 1901.
Baddeley, Alan. "Working Memory." *Science* 255 (1992) 556–59.
Bak, Per. *How Nature Works: The Science of Self-Organized Criticality*. New York: Springer, 1999.
Barbour, Ian G. *Religion and Science: Historical and Contemporary Issues*. New York: HarperOne, 1997.
Barclay, William. *The Gospel of John, Volume 1*. Philadelphia: Westminster, 1975.
———. *The Mind of Jesus*. New York: Harper & Row, 1961.
Barfield, Owen. *Saving the Appearances: A Study in Idolatry*. Middletown, CT: Wesleyan University Press, 1988.
Barker, Joel A. *Discovering the Future: The Business of Paradigms*. St. Paul, MN: ILI, 1989.
Baron, Robert J. *The Cerebral Computer: An Introduction To the Computational Structure of the Human Brain*. Hillsdale, NJ: Lawrence Erlbaum, 1987.
Barr, Stephen M. *Modern Physics and Ancient Faith*. Notre Dame: University of Notre Dame Press, 2006.
Bartholomew, David J. *God of Chance*. London: SCM, 1984.
Bechara, Antoine, et al. "Emotion, Decision Making, and the Orbitofrontal Cortex." *Cerebral Cortex* 10 (2000) 295–307.

Bechtel, William, and George Graham, eds. *A Companion to Cognitive Science*. Malden, MA: Blackwell, 1999.
Bell, E. T. *Men of Mathematics: The Lives and Achievements of the Great Mathematicians from Zeno to Poincaré*. New York: Simon & Schuster, 1986.
Benchley, Robert C. *Of All Things*. New York: Henry Holt, 1921.
Berlinski, David. *The Devil's Delusion: Atheism and Its Scientific Pretensions*. New York: Basic, 2009.
Berry, Wendell. *Life is a Miracle: An Essay Against Modern Superstition*. Berkeley: Counterpoint, 2001.
Bickerton, Derek. *Language and Human Behavior*. Seattle: University of Washington Press, 1995.
Bierce, Ambrose. *The Devil's Dictionary*. New York: Dover, 1958.
Billings, Josh. *Everybody's Friend, or; Josh Billing's Encyclopedia and Proverbial Philosophy of Wit and Humor*. Hartford: American, 1874.
———. *Josh Billings, His Sayings*. New York: Carleton, 1865.
Bowles, Colin, ed. *The Wit's Dictionary*. London: Angus & Robertson, 1984.
Boyd, Brian. "Purpose-Driven Life." *The American Scholar* 78.2 (2009) 24–34.
Browne, Malcolm. "Physicists Debunk Claims of a New Kind of Fusion." *The New York Times on the Web*, May 3, 1989. http://partners.nytimes.com/library/national/science/050399sci-cold-fusion.html.
Buckingham, Lindsey. "Go Your Own Way." *Rumours*. Fleetwood Mac. Burbank: Warner Brothers Records, 1977.
Bunch, Wilton H. "Two Stories are Better than One: Looking Through the Lens of Faith and Science." *Spectrum* 39.3 (2011) 24–28.
Bunyan, John. *The Pilgrim's Progress*. Springdale, PA: Whitaker House, 1973.
Burnham, Tom. *The Dictionary of Misinformation*. New York: Ballantine, 1977.
Bustamante, Homero. "Concept of Faith an Intellectual Cop-out." *The Birmingham News*. January 19, 2007.
Byrne, Peter. "The Many Worlds of Hugh Everett." *Scientific American* 297.6 (2007) 98–105.
Byrne, Richard. *The Thinking Ape: Evolutionary Origins of Intelligence*. Oxford: Oxford University Press, 1995.
Cappon, Lester J., ed. *The Adams-Jefferson Letters*. Chapel Hill, NC: University of North Carolina Press, 1988.
Carus, T. Lucretius. *On the Nature of Things*. Translated by William Leonard. Charleston, SC: Forgotten Books, 2007.
Carlyle, Thomas. *Oliver Cromwell's Letters and Speeches, Volume 1*. New York: Harper & Brothers, 1855.
Cerf, Vinton. "Google's View." *Smithsonian* 41.4 (2010) 120.
Chabris, Christopher and Daniel Simon. *The Invisible Gorilla: How Our Intuitions Deceive Us*. New York: Broadway, 2011.
Charlesworth, Brian and Deborah Charlesworth. *Evolution: A Very Short Introduction*. New York: Oxford University Press, 2003.
Cicero, Marcus. *De Natura Deorum*. Translated by H. Rackham. Cambridge, MA: Harvard University Press, 1933.
———. *The Nature of the Gods*. Translated by P. G. Walsh. Oxford: Clarendon, 1997.
Clarke, Arthur C. *Profiles of the Future: An Inquiry into the Limits of the Possible*. New York: Holt, Rinehart & Winston, 1984.

Cory, Lloyd. *Quotable Quotations*. Wheaton, IL: Victor, 1985.
Crick, Francis. *The Astonishing Hypothesis: The Scientific Search for the Soul*. New York: Simon & Schuster, 1994.
———. *Life Itself: Its Origin and Nature*. New York: Simon & Schuster, 1981.
Crumbo, Kim. *A River Runner's Guide to the History of the Grand Canyon*. Boulder, CO: Johnson, 1981.
Damasio, Antonio. *Descartes' Error: Emotion, Reason, and the Human Brain*. New York: Avon, 1995.
Darwin, Charles. *On Origin of Species*. London: Penguin, 1982.
Davies, Paul. *The Mind of God*. New York: Simon & Schuster, 1993.
Dawkins, Richard. *A Devil's Chaplain: Reflections on Hope, Lies, Science, and Love*. Boston: Houghton Mifflin, 2003.
———. *The God Delusion*. Boston: Houghton Mifflin, 2006.
———. "The Illusion of Design." *Natural History*. November 2005. http://www.naturalhistorymag.com/htmlsite/master.html?http://www.naturalhistorymag.com/htmlsite/1105/1105_feature1.html.
Dennett, Daniel C. *Kinds of Minds: Toward An Understanding Of Consciousness*. New York: Basic, 1997.
———. *Sweet Dreams: Philosophical Obstacles to a Science of Consciousness*. Cambridge, MA: MIT Press, 2005.
Dennett, Daniel C., and Alvin Plantinga. *Science and Religion: Are They Compatible?* New York: Oxford University Press, 2011.
Desimone, Robert. "The Physiology of Memory: Recordings of Things Past." *Science* 258 (1992) 245–46.
DeVinne, Pamela B., ed. *The American Heritage Dictionary*. Boston: Houghton Mifflin, 1991.
Diamond, Jared. *Collapse: How Societies Choose to Fail or Succeed*. New York: Viking, 2005.
Dilday, Russell H. *The Doctrine of Biblical Authority*. Nashville: Convention, 1982.
Disraeli, Benjamin. *Coningsby; Or, the New Generation*. Volume 2. London: Henry Colburn, 1844.
———. *Tancred: Or the New Crusade*. London: John Lane, 1905.
Donaldson, Steve. "A Neural Network for Creative Serial Order Cognitive Behavior." *Minds and Machines* 18 (2008) 53–91.
———. "Predictive Learning and Cognitive Momentum: A Foundation for Intelligent, Autonomous Systems." *Proceedings of the 37th Southeast Regional Conference of the ACM*, 1999.
Drummond, Henry. *The Lowell Lectures on the Ascent of Man*. Radford, VA: Wilder, 2008.
Edelman, Gerald M. *Bright Air, Brilliant Fire: On the Matter of the Mind*. New York: Basic, 1992.
Eiland, Morgan. "By Definition, Faith Not Based on Proof." *The Birmingham News*. January 20, 2007.
Fauconnier, Gilles, and Mark Turner. *The Way We Think: Conceptual Blending and the Mind's Hidden Complexities*. New York: Basic, 2003.
Feldman. Michael. "Must See TV: IBM Watson Heads for Jeopardy Showdown." *HPCwire*, February 9, 2011. http://www.hpcwire.com/2011/02/09/must_see_tv_ibm_watson_heads_for_jeopardy_showdown/.

Ferris, Timothy. *The Science of Liberty: Democracy, Reason, and the Laws of Nature.* New York: Harper, 2010.
Flew, Antony, and Roy A. Varghese. *There is a God: How the World's Most Notorious Atheist Changed His Mind.* New York: HarperOne, 2008.
Friend, Tad. "Jumpers: The Fatal Grandeur of the Golden Gate Bridge." *The New Yorker* October 13, 2003. http://www.newyorker.com/magazine/2003/10/13/jumpers.
Galilei, Galileo. *Dialogues Concerning Two New Sciences.* Translated by Henry Crew and Alfonso De Salvio. New York: Macmillan, 1914.
———. *Dialogue Concerning the Two Chief World Systems.* Translated by Stillman Drake. New York: Modern Library, 2001.
Garreau, Joel. *Radical Evolution.* New York: Broadway, 2006.
Garside, Rick, director. *Faith Happens.* Side by Side Films, 2006.
Gersting, Judith L. *Mathematical Structures for Computer Science.* New York: W. H. Freeman, 2007.
Giberson, Karl. *Saving Darwin: How to Be a Christian and Believe in Evolution.* New York: HarperOne, 2008.
Gilliam, Terry, and Terry Jones, directors. *Monty Python and the Holy Grail.* Michael White Productions, 1974.
Gilovich, Thomas. *How We Know What Isn't So: The Fallibility of Human Reason in Everyday Life.* New York: Free Press, 1991.
Gilovich, Thomas, et al. "The Hot Hand in Basketball: On the Misperception of Random Sequences." *Cognitive Psychology* 17 (1985) 295–314.
Girard, G. "The Third Periodic Verification of National Prototypes of the Kilogram." *Metrologia* 31 (1994) 317–36.
Gödel, Kurt. "Über formal unentscheidbare Sätze der Principia Mathematica und verwandter Systeme, I." *Monatshefte für Mathematik und Physik* 38 (1931) 173–98.
Goldsmith, Oliver. *The Deserted Village.* London: W. Griffin, 1770.
Gould, Stephen J. "Nonoverlapping Magisteria." *Natural History* 106 (1997) 16–22.
Green, Joel. *Body, Soul, and Human Life: The Nature of Humanity in the Bible.* Grand Rapids: Baker Academic, 2008.
Grogan, Barbara B., ed. *The Knowledge Book: Everything You Need to Know to Get by in the 21st Century.* Washington, DC: National Geographic, 2009.
Hakel, Milton D. "How Often is Often?" *American Psychologist* 23 (1968) 533–34.
Harris, Sam. *The End of Faith: Religion, Terror, and the Future of Reason.* New York: W. W. Norton, 2005.
Harrison, Peter. *The Bible, Protestantism, and the Rise of Natural Science.* Cambridge: Cambridge University Press, 2001.
———, ed. *The Cambridge Companion to Science and Religion.* Cambridge: Cambridge University Press, 2010.
Haught, John F. *Christianity and Science: Toward a Theology of Nature.* Maryknoll, NY: Orbis, 2007.
Hawking, Stephen. *A Brief History of Time.* New York: Bantam, 1998.
Hawkins, Jeff, and Sandra Blakeslee. *On Intelligence.* New York: St. Martin's Griffin, 2004.
Henig, Robin M. "Darwin's God." *The New York Times*, March 4, 2007. http://www.nytimes.com/2007/03/04/magazine/04evolution.t.html?pagewanted=all&_r=0.
Heisenberg, Werner. *Physics and Beyond.* Translated by Arnold J. Pomerans. New York: Harper & Row, 1971.

Herculano-Houzel, Suzana. "The Human Brain in Numbers: A Linearly Scaled-up Primate Brain." *Frontiers in Human Neuroscience* 3:31 (2009).
Hofstadter, Douglas R. *Gödel, Escher, Bach: An Eternal Golden Braid*. New York: Basic, 1979.
———. *I Am a Strange Loop*. New York: Basic, 2007.
———. *Metamagical Themas: Questing for the Essence of Mind and Pattern*. New York: Basic, 1985.
Horgan, John. *The End of Science: Facing the Limits of Knowledge in the Twilight of the Scientific Age*. Reading, MA: Addison-Wesley, 1996.
———. "Why I Think Science is Ending." *Edge*, May 6, 1997. http://www.edge.org/documents/archive/edge16.html.
Hull, Edward. *The Wall Chart of World History: From Earliest Times to the Present*. London: BestSeller, 1988.
Hume, David. *Philosophical Essays Concerning Human Understanding*. London: A. Millar, 1748.
Huxley, Thomas H. "On the Advisableness of Improving Natural Knowledge." *Fortnightly Review* 3 (1866) 626–37.
Isaacson, Walter. *Einstein: His Life and Universe*. New York: Simon & Schuster, 2007.
Jackendoff, Ray S. *Consciousness and the Computational Mind*. Cambridge, MA: MIT Press, 1987.
James, William. *The Varieties of Religious Experience*. Scotts Valley, CA: IAP, 2009.
———. *The Will to Believe & Other Essays*. Seaside, OR: Watchmaker, 2010.
Jastrow, Robert, and Michael Rampino. *Origins of Life in the Universe*. New York: Cambridge University Press, 2009.
Jaynes, Julian. *The Origin of Consciousness in the Breakdown of the Bicameral Mind*. Boston: Houghton Mifflin, 1976.
Jefferson, Thomas. "Letter to William Roscoe." Library of Congress Archives, December 27, 1820. http://www.loc.gov/exhibits/jefferson/75.html.
Jeffress, Lloyd A., ed. *Cerebral Mechanisms in Behavior*. New York: John Wiley, 1951.
Jenkins, Mark. "The Good Company of the Dead." *Outside Magazine*, August 1999. http://www.thehardway.com/stories/mallory.htm.
Jung, Carl. "Face to Face." BBC interview. Produced by Hugh Burnett, 1959.
Kandel, Eric, et al., eds. *Essentials of Neural Science and Behavior*. Norwalk, CT: Appleton & Lange, 1995.
Kauffman, Stuart. *At Home in the Universe: The Search for the Laws of Self-Organization and Complexity*. New York: Oxford University Press, 1996.
———. *Reinventing the Sacred: A New View of Science, Reason and Religion*. New York: Basic, 2008.
Kaufman, Gordon D. *Jesus and Creativity*. Minneapolis: Fortress, 2006.
King, Martin. L., Jr. *Strength to Love*. Minneapolis: Fortress, 2010.
Kitcher, Philip. "The Fact of Evolution." *The New York Times*, October 23, 2009. http://www.nytimes.com/2009/10/25/books/review/Letters-t-THEFACTOFEVO_LETTERS.html?_r=0.
Kluger, Jeffrey, et al. "Is God in Our Genes?" *Time* 164.17 (2004) 62–72.
Koch, Christof, and Joel L. Davis, eds. *Large-Scale Neuronal Theories of the Brain*. Cambridge, MA: MIT Press, 1994.
Krauss, Lawrence M. "God and Science Don't Mix." *The Wall Street Journal*, June 26, 2009. http://www.wsj.com/articles/SB124597314928257169.

Krauthammer, Charles, and W. Dowell. "Deep Blue Funk." *Time* 147.9 (1996) 60.
Krell, Dorothy, ed. *National Parks of the West*. Menlo Park, CA: Lane, 1980.
Kuhl, Patricia K. "Learning and Representation in Speech and Language." *Current Opinion in Neurobiology* 4 (1994) 812–22.
Kuhn, Robert L., ed. *Closer to Truth: Challenging Current Belief*. New York: McGraw-Hill, 2000.
Kuhn, Thomas S. "The Essential Tension." In *The Third University of Utah Research Conference on the Identification of Scientific Talent*, edited by C. W. Taylor, 21–30. Salt Lake City: University of Utah Press, 1959.
———. *The Structure of Scientific Revolutions*. Chicago: University of Chicago Press, 1996.
Kurtz, Paul. *Science and Religion: Are They Compatible?* New York: Prometheus, 2003.
Lakoff, George. *Women, Fire, and Dangerous Things: What Categories Reveal About the Mind*. Chicago: University of Chicago Press, 1987.
Lanza, Michael. "Speed Freak." *Backpacker* 32.4 (2004) 78–119.
Laplace, Pierre S. *A Philosophical Essay on Probabilities*. Translated by Frederick W. Truscott and Frederick L. Emory. New York: John Wiley & Sons, 1902.
Latour, Bruno, and Steve Woolgar. *Laboratory Life: The Construction of Scientific Facts*. Princeton: Princeton University Press, 1986.
Lewis, C. S. *The Discarded Image: An Introduction to Medieval and Renaissance Literature*. Cambridge: Cambridge University Press, 1964.
———. *Mere Christianity*. Westwood, NJ: Barbour, 1952.
———. *Surprised by Joy*. San Diego: Harcourt Brace Jovanovich, 1956.
Libet, Benjamin, et al., eds. *The Volitional Brain: Towards a Neuroscience of Free Will*. Exeter, UK: Imprint Academic, 1999.
Lienhard, John H. "No. 1233: Baden Baden-Powell." In *Engines of Our Ingenuity*. www.uh.edu/engines/epi1233.htm.
Lindley, David. *The End of Physics: The Myth of a Unified Theory*. New York: Basic, 1994.
Lindsey, Hal. *The Late Great Planet Earth*. Grand Rapids: Zondervan, 1970.
Lloyd, Seth. *Programming the Universe: A Quantum Computer Scientist Takes on the Cosmos*. New York: Vintage, 2007.
Locke, John. *An Essay Concerning Human Understanding*. In *Great Books of the Western World*, edited by Robert Maynard Hutchins. Chicago: William Benton, 1952.
Lorentz, Edward N. "Deterministic Nonperiodic Flow." *Journal of the Atmospheric Sciences* 20 (1963) 130–41.
Luger, George F. *Artificial Intelligence: Structures and Strategies for Complex Problem Solving*. New York: Addison-Wesley, 2005.
Mace, C. A., ed. *British Philosophy in Mid-Century*. London: George Allen and Unwin, 1957.
MacPhail, Euan. *The Evolution of Consciousness*. Oxford: Oxford University Press, 1998.
Maritain, Jacques. *An Introduction to Philosophy*. Westminster, MD: Christian Classics, 1989.
Maxwell, John C., and Jim Dornan. *Becoming a Person of Influence: How to Positively Impact the Lives of Others*. Nashville: Thomas Nelson, 2006.
McClay, Wilfred M. "Revisiting the Idea of Progress in History." *Historically Speaking* 9.1 (2007) 11–12.
McClelland, James L., and David E. Rumelhart. *Explorations in Parallel Distributed Processing*. Cambridge, MA: MIT Press, 1988.

McCulloch, Warren S., and Walter Pitts. "A Logical Calculus of the Ideas Immanent in Nervous Activity." *Bulletin of Mathematical Biophysics* 5 (1943) 115–33.
McGilchrist, Iain. *The Master and His Emissary: The Divided Brain and the Making of the Western World*. New Haven: Yale University Press, 2012.
McGrath, Alister. *Surprised by Meaning: Science, Faith, and How We Make Sense of Things*. Louisville: Westminster John Knox, 2011.
McNamara, Thomas E. *Evolution, Culture, and Consciousness: The Discovery of the Preconscious Mind*. Lanham, MD: University Press of America, 2004.
McNeill, Daniel, and Paul Freiberger. *Fuzzy Logic*. New York: Simon & Schuster, 1993.
Meldahl, Keith H. *Rough-Hewn Land: A Geologic Journey from California to the Rocky Mountains*. Berkeley: University of California Press, 2011.
Mencken, H. L. *A Mencken Chrestomathy: His Own Selection of His Choicest Writing*. New York: Vintage, 1982.
———. *Prejudices, Third Series*. New York: Alfred A. Knopf, 1922.
Mill, John Stuart. *On Liberty*. Ontario: Batoche, 2001.
Miller, George A. "The Magical Number Seven, Plus or Minus Two: Some Limits on Our Capacity for Processing Information." *Psychological Review* 63 (1956) 81–97.
Miller, Perry, ed. *The American Puritans: Their Prose and Poetry*. Garden City, NY: Doubleday Anchor, 1956.
Miner, Horace. "Body Ritual Among the Nacirema." *American Anthropologist* 58 (1956) 503–7.
Minsky, Marvin. *The Society of Mind*. New York: Simon and Schuster, 1985.
Mlodinow, Leonard. *The Drunkard's Walk: How Randomness Rules Our Lives*. New York: Pantheon, 2008.
Montaigne, Michel de. *The Complete Essays of Montaigne*. Translated by Donald M. Frame. Stanford: Stanford University Press, 1958.
Mooney, Chris and Sheril Kirshenbaum. "Why America is Flunking Science." *Salon*, July 13, 2009. http://www.salon.com/2009/07/13/science_illiteracy/.
Moravec, Hans. *Robot: Mere Machine to Transcendent Mind*. New York: Oxford University Press, 1999.
———. "When Will Computer Hardware Match the Human Brain?" *Journal of Evolution and Technology*, Vol. 1 (1998) http://www.jetpress.org/volume1/moravec.pdf.
Morowitz, Harold J. *The Emergence of Everything: How the World Became Complex*. New York: Oxford University Press, 2002.
Morris, Simon Conway, ed. *The Deep Structure of Biology*. West Conshohocken, PA: Templeton Foundation, 2008.
Naam, Ramez. *More Than Human*. New York: Broadway, 2005.
Nietzsche, Friedrich. *The Anti-Christ*. New York: SoHo, 2012.
———. *Beyond Good and Evil: Prelude to a Philosophy of the Future*. Translated by Walter Kaufmann. New York: Vintage, 1966.
Numbers, Ronald L., ed. *Galileo Goes to Jail and Other Myths About Science and Religion*. Cambridge, MA: Harvard University Press, 2010.
Ortenburger, Leigh N., and Reynold G. Jackson, R. *A Climber's Guide to the Teton Range*. Seattle: The Mountaineers, 1996.
Panek, Richard. *Seeing and Believing: How the Telescope Opened Our Eyes and Minds to the Heavens*. New York: Penguin, 1998.

Parker-Pope, Tara. "Mountain Climbing Bad for the Brain." *The New York Times*, October 20, 2008. http://well.blogs.nytimes.com/2008/10/20/mountain-climbing-bad-for-the-brain/.
Pascal, Blaise. *Pensées*. Translated by W. F. Trotter. Great Books of the Western World. Chicago: William Benton, 1952.
Peacocke, Arthur. *Creation and the World of Science: The Re-Shaping of Belief*. New York: Oxford University Press, 2004.
———. *Intimations of Reality: Critical Realism in Science and Religion*. Notre Dame: University of Notre Dame Press, 1984.
Penrose, Roger. *The Emperor's New Mind: Concerning Computers, Minds and The Laws of Physics*. Oxford: Oxford University Press, 1989.
Peterson, Gregory R. *Minding God: Theology and the Cognitive Sciences*. Minneapolis: Fortress, 2003.
Pillemer, David B. "Clarifying the Flashbulb Memory Concept: Comment on McCloskey, Wible, and Cohen." *Journal of Experimental Psychology: General* 119.1 (1990) 92–96.
Pinker, Steven. *How the Mind Works*. New York: W. W. Norton, 1997.
———. *The Language Instinct: How the Mind Creates Language*. New York: Harper Perennial Classics, 2000.
Pirsig, Robert M. *Zen and the Art of Motorcycle Maintenance: An Inquiry Into Values*. New York: HarperTorch, 2006.
Planck, Max. *Where Is Science Going?* Woodbridge, CT: Oxbow, 1981.
Plato. *Gorgias*. Translated by James H. Nichols, Jr. Ithaca, NY: Cornell University Press, 1998.
———. *Meno*. Translated by Benjamin Jowett. Stilwell, KS: Digireads, 2005.
Poe, Edgar Allan. "The Raven." In *Great Tales and Poems of Edgar Allan Poe*. New York: Pocket, 2007.
Polanyi, Michael. *Personal Knowledge: Towards a Post-Critical Philosophy*. Chicago: University of Chicago Press, 1962.
Polkinghorne, John, and Nicholas Beale. *Questions of Truth: Fifty-one Responses to Questions About God, Science, and Belief*. Louisville: Westminster John Knox, 2009.
Post, Ted, director. *Magnum Force*. Warner Brothers, 1973.
Powell, John Wesley. *Exploration of the Colorado River and Its Canyons*. Mineola, NY: Dover, 1961.
Prigatano, George P., and Daniel L. Schacter, eds. *Awareness of Deficit After Brain Injury*, Oxford: Oxford University Press, 1991.
Proust, Marcel. *In Search of Lost Time: The Captive*. Translated by C. K. Scott Moncrieff, et al. New York: Modern Library, 2003.
Purves, Dale, and R. Beau Lotto. *Why We See What We Do: An Empirical Theory of Vision*. Sunderland, MA: Sinauer, 2003.
Radowitz, John von. "Back to the Beginning: Hadron Collider Creates Mini-Big Bang." *The Independent*, November 9, 2010. http://www.independent.co.uk/news/science/back-to-the-beginning-hadron-collider-creates-minibig-bang-2128785.html
Reiner, Carl, director. *Oh, God!* Warner Brothers, 1977.
Reiner, Rob, director. *A Few Good Men*. Castle Rock Entertainment, 1992.

Reppert, Victor. *C. S. Lewis's Dangerous Idea: In Defense of the Argument from Reason.* Downers Grove, IL: InterVarsity, 2003.
Roach, Gerry. *Colorado's Fourteeners: From Hikes to Climbs.* Golden, CO: Fulcrum, 1999.
Robbins, Ray Frank. *The Revelation of Jesus Christ.* Nashville: Broadman, 1975.
Robinson, Marilynne. *Absence of Mind: The Dispelling of Inwardness from the Modern Myth of the Self.* New Haven: Yale University Press, 2010.
Roco, Mihail C., and Williams Sims Bainbridge, eds. *Converging Technologies for Improving Human Performance: Nanotechnology, Biotechnology, Information Technology and Cognitive Science.* Dordrecht, Netherlands: Kluwer Academic, 2003.
Rolston III, Holmes. *Genes, Genesis, and God: Values and Their Origins in Natural and Human History.* Cambridge: Cambridge University Press, 1999.
Rose, Steven, ed. *From Brains to Consciousness? Essays on the New Sciences of the Mind.* Princeton: Princeton University Press, 1998.
Ross, Hugh. *More Than a Theory: Revealing a Testable Model for Creation.* Grand Rapids: Baker, 2009.
Rowell, Galen. *Mountain Light: In Search of the Dynamic Landscape.* San Francisco: Sierra Club, 1995.
Ruffin, Steven A. *Aviation's Most Wanted: The Top 10 Book of Winged Wonders, Lucky Landings, and Other Aerial Oddities.* Washington, D.C.: Potomac, 2005.
Sacks, Oliver. *The Man Who Mistook His Wife for a Hat (and Other Clinical Tales).* New York: Touchstone, 1998.
Sagan, Carl. *Cosmos.* New York: Random House, 1980.
Sartre, Jean-Paul. *Existentialism and Human Emotions.* New York: Philosophical Library, 1957.
Saxe, John Godfrey. *The Poems of John Godfrey Saxe.* Boston: James R. Osgood, 1873.
Schacter, Daniel L. *Searching for Memory: The Brain, The Mind, And The Past.* New York: Basic, 1997.
Schneider, Susan, ed. *Science Fiction and Philosophy: From Time Travel to Superintelligence.* Malden, MA: Wiley-Blackwell, 2009.
Schrödinger. Erwin. *What Is Life? with "Mind and Matter" and "Autobiographical Sketches."* Cambridge: Cambridge University Press, 2006.
Schults, W., and K. Swan. "Retinal Hemorrhages in Mountain Climbers." *Summit* 21.2 (1975) 19–26.
Schumacher, Joel, director. *Batman Forever.* Warner Brothers, 1995.
Searle, John R. "Minds, Brains, and Programs." *Behavioral and Brain Sciences* 3 (1980) 417–57.
Shakespeare, William. *The Merchant of Venice.* New York: Dover, 1995.
Shannon, Claude E. "Programming a Computer for Playing Chess." *Philosophical Magazine* 41.314 (1950) 256–75.
Shermer, Michael. *The Science of Good & Evil: Why People Cheat, Gossip, Care, Share, and Follow the Golden Rule.* New York: Henry Holt, 2004.
Simonton, Dean Keith. *Creativity in Science: Chance, Logic, Genius, and Zeitgeist.* New York: Cambridge University Press, 2004.
Simpson, Ray H. "The Specific Meanings of Certain Terms Indicating Differing Degrees of Frequency." *The Quarterly Journal of Speech* 30 (1944) 328–30.
Smith, Christian. *The Bible Made Impossible.* Grand Rapids: Brazos, 2011.

Smolin, Lee. *The Life of the Cosmos*. New York: Oxford University Press, 1997.
Soon, Chun Siong, et al. "Unconscious Determinants of Free Decisions in the Human Brain." *Nature Neuroscience* 11 (2008) 543–5.
Spielberg, Nathan, and Byron D. Anderson. *Seven Ideas That Shook the Universe*. New York: John Wiley & Sons, 1995.
Spitzer, Robert J. *New Proofs for the Existence of God*. Grand Rapids: Eerdmans, 2010.
Stannard, Russell. *The End of Discovery: Are We Approaching the Boundaries of the Knowable?* New York: Oxford University Press, 2010.
Stravinsky, Igor. *Poetics of Music in the Form of Six Lessons*. Cambridge: Harvard University Press, 1993.
Strogatz, Steven H. *Sync: How Order Emerges From Chaos In the Universe, Nature, and Daily Life*. New York: Hyperion, 2003.
Swinburne, Richard. *The Concept of Miracle*. London: Macmillan, 1970.
———. *Is There a God?* Oxford: Oxford University Press, 1996.
Taleb, Nassim Nicholas. *The Black Swan: The Impact of the Highly Improbable*. New York: Random House, 2007.
Taunton, Larry. "My Take: An evangelical remembers his friend Hitchens." *CNN Belief Blog*, December 16, 2011. http://religion.blogs.cnn.com/2011/12/16/my-take-an-evangelical-remembers-his-friend-hitchens/
Taylor, Barbara. *The Luminous Web: Essays on Science and Religion*. Cambridge, MA: Cowley, 2000.
Teilhard de Chardin, Pierre. *The Phenomenon of Man*, New York: Harper Perennial, 2008.
Thomas, Cal. "Hitchens Smart, But Falls Short of Being Wise." *The Birmingham News*. December 22, 2011. 7A.
Thorndike, Lynn. *A History of Magic and Experimental Science During the First Thirteen Centuries of Our Era*. New York: Columbia University Press, 1923.
Tillich, Paul. *Dynamics of Faith*. New York: Harper & Row, 1958.
Turing, A. M. "Computing Machinery and Intelligence." *Mind* 59 (1950) 433–60.
Tversky, Amos, and Daniel Kahneman. "The Framing of Decision and the Psychology of Choice." *Science* 211 (1981) 453–58.
———. "Judgment Under Uncertainty: Heuristics and Biases." *Science* 185 (1974) 1124–31.
Twain, Mark. *Life on the Mississippi*. New York: Harper & Brothers, 1901.
Velmans, Max, and Susan Schneider, eds. *The Blackwell Companion to Consciousness*. Malden, MA: Blackwell, 2007.
Vos Savant, Marilyn. "Ask Marilyn." *Parade Magazine*. September 22, 1996.
———. "Ask Marilyn." *Parade Magazine*. October 8, 2000.
Walker, Evan Harris. *The Physics of Consciousness: The Quantum Mind And The Meaning Of Life*. Cambridge, MA: Perseus, 2000.
Ward, Keith. *The Big Questions in Science and Religion*. West Conshohocken, PA: Templeton Foundation, 2008.
———. *Divine Action: Examining God's Role in an Open and Emergent Universe*. West Conshohocken, PA: Templeton Foundation, 2007.
Watts, Fraser, ed. *Creation: Law and Probability*. Minneapolis: Fortress, 2008.
Weil, Simone. *Waiting on God*. New York: Routledge, 2010.
Weinberg, Gerald M. *An Introduction to General Systems Thinking*. New York: Dorset House, 2001.

Wilder, Thornton. *Our Town*. New York: Harper, 2003.
Wilson, Brian, and Mike Love. *Good Vibrations*. Beach Boys. Los Angeles: Capitol Records, 1966.
Wilson, Edward O. *Consilience: The Unity of Knowledge*. New York: Vintage, 1999.
———. *The Creation: An Appeal to Save Live on Earth*. New York: W. W. Norton, 2007.
Winokur, Jon, ed. *The Portable Curmudgeon*. New York: Plume, 1992.
Wolfram, Stephen. *A New Kind of Science*. Champaign, IL: Wolfram Media, 2002.
Yerxa, Donald A. *Recent Themes in the History of Science and Religion*. University of South Columbia, SC: Carolina Press, 2009.
Young, Simon. *Designer Evolution: A Transhumanist Manifesto*. Amherst, NY: Prometheus, 2006.
Zacharias, Ravi. *Cries of the Heart*. Nashville: Thomas Nelson, 2002.
Zadeh, Lofti A. "Fuzzy Logic = Computing with Words." *IEEE Transactions on Fuzzy Systems* 4.2 (1996) 103–11.

Photo Credits

Chapter 1, Figure 1: Steve Donaldson
Chapter 3, Figure 1: Steve Donaldson
Chapter 6, Figure 1: Carol Donaldson

Subject Index

51 Pegasi, 134–35
Abbott, Edwin A., 81n8
absolute zero, 97n36
Aconcagua, 81
Adam, 67, 83
Adams, John, 188
Adams, Scott, 47n12
afterlife, 177, 225. *See also* heaven.
agnostic, 101
algorithm, 104, 159n35
Albert the Great, 35n3
Albus, James S., 71n47
Alzheimer's, 75, 90
amnesia, retrograde, 90
Anasazi, 140n5
Anderson, Byron D., 110n78, 110n80, 111n83
Anderson, Chris, 162n41
anosognosia, 82, 83n14, 89, 175n5
answers
    as approximations, 90–95
    believed impossible to find, 190
    better, 175n7
    to big questions, xii, 141, 161, 244–45
    conflicting, 164
    in Factland, 240–42
    generating new questions, 111n83
    and God's clarity, 202, 207, 212–13, 214n21, 216, 218
    to key questions about miracles, 178
    rational, 234n45
    to two fundamental types of questions, 36
    variation in, 246
Anthony, Susan B., 148
anthropologist, 100
anthropology, 159
anthropomorphic, 172n2, 227
Anton's syndrome, 175n5
apologetics, 191
apostasy, 191
approximations
    and answers to significant questions, 90–95
    and measurements, 95–100
    and theories, 100–104
Aquinas, Thomas, 35n3, 57n9, 148
Arches National Park, 5
Archimedes, 16n6
Aristotle, 31, 102–3, 126–28, 134, 139, 155
armchair quarterback, 159
arrogance, xii, 163–64, 173, 232n36. *See also* pride.
Arctic Circle, 18
artificial intelligence, xiii, 70, 70n47
assumption(s)
    and conflicting beliefs, 28
    defense of, 3–6
    and fact-only world, 241
    about faith, 31, 125
    about God, 203, 207–8, 212
    about "hot hand" in basketball, 86–87
    invalid, 88–89
    limiting understanding, 188
    and making good choices, xii

assumption(s) (continued)
  about marriage, 36
  Nietzsche's, 18
  origins of, xv
  about Pascal's wager, 235
  about physical explanation for mental properties, 70
  and relation to memory, 89–90
  about rigidity of religious faith, 197
  and role in crafting rational faith, 133–35
  about science as religion, 199
  about the supernatural, 187
  about time for life to arise on earth, 198
  about truth seekers, 247
  See also premise.
atheism, 208, 245–46
atheist, xiii, 9, 15n5, 27, 101, 129, 173, 180, 204–7, 213, 218–20, 229n21, 232, 235, 245–46
Atran, Scott, 172n3
attention
  deficits, 62–63
  directed by explorers, 156
  drawn to God, 215
  to the trivial, 202, 210
attitude
  toward evidence, 146, 205, 213
  of explorers, 153–5
  inhibiting development of rational faith, 159–60, 163, 166
  toward miracles, 180, 224
  toward the soul, 73
  toward worship, 194
Augustine, Saint, 148, 156
authority, 44, 134–35, 158, 165–66, 181–82, 224n3
automaton, 39, 243
autonomy, 243n6
awareness, 34, 58, 67n34, 81–83, 89, 105n64, 175n5, 243n4

Baddeley, Alan, 56n2, 81n7
Bacon, Francis, 85, 126, 129n38, 135, 161n40, 214n21, 245–46
Bainbridge, William Sims, 81n9
Bak, Per, 198n26, 250n26

Baldwin effect, 10
Barbour, Ian G., 69n40, 104n60, 122n23, 197n20
Barclay, William, 185–86
Barfield, Owen, 198
Barker, Joel A, 122n22
Baron, Robert J., 56n2
Barr, Stephen M., 44n10, 74–75
Bartholomew, David J., 47n13, 214n21
Batman, 42n8
Beach Boys, 227–28
Bechara, Antoine, 70n44
behavior
  adaptive, 63–70
  affected by God, 216
  of deity, 177, 180, 205, 225
  engendered by belief, 90, 113–14
  explorer's, 155–60
  influenced by environment, 121
  irrational, 31
  naïve, 176–77
  neuron, 59, 71n47
  reductionist view of, 196
  religious, 193, 198
  restricted by cerebral deficits, 82
  stereotyped, 67n35
behaviorism, 122
belief. *See* faith.
belief cycle, 179–80. *See also* epistemic circle.
belief probability, 136–37, 234
belief trajectory, 60–61, 225, 246–47
believer, 14, 172, 181
Bell, E. T., 185n24
Bell Labs, 93
Belousov, Boris, 103, 106
Belousov-Zhabotinsky reaction, 103
Belshazzar, 99
Benchley, Robert C., 139–40
Bentall, Richard, 124n27
Berlinski, David, 224n7
Berry, Wendell, 14–15, 53n23, 163n42, 243n3
Bible, 22, 166n53, 182, 232n36. *See also* sacred text.
biblical, 26, 83, 99, 129, 157n29, 190, 215, 236n55
Bickerton, Derek, 66n32

SUBJECT INDEX

Bierce, Ambrose, 14–18, 23, 132, 150n16
Big Bang, 198
Billings, Josh, 4, 30, 144
biologist, 101–2, 129, 194, 235. *See also* evolutionary biologist.
biology, 159, 214. *See also* evolutionary biology.
Black Canyon, 229
black hole, 198, 220n35
Blakeslee, Sandra, 64n25, 167n55
Bliss, Tim, 59n14
blind faith. *See* faith, blind.
Bowles, Colin, 117n10
Boyd, Brian, 145n12, 193–94
brain
   and behavior, 55–76, 79, 216
   and faith, xv, 4, 55–76, 176, 228
   internal connectivity of, 10, 10n6
   and learning, 157–9
   limitations of, 80–90, 94, 104
   as a probability machine, 57–63
   and the soul, 70–76
   and what one knows for sure, 43
Bryce Canyon National Park, 153
Bryce, Ebenezer, 153
Buddha, Gautama, 213
Buddhism, 174, 194, 213
Bunch, Wilton, 29, 111n82, 178n12
Bunyan, John, 153
Bureau International des Poids et Mesures, 97–98
Burnham, Tom, 12n10
Burns, George, 204
Bustamante, Homero, 3
butterfly effect, cognitive, 62
Byrne, Peter, 95n31
Byrne, Richard, 66n30,

cairn, xi–xiii
Calvinism and faith, 119n19
Canyonlands National Park, 195
Captain Kirk, 141
cardio-pulmonary resuscitation, 209
categorization, 15, 18, 57n6, 140. *See also* classification.
causation, 162n41
Cerf, Vinton, 165n45

certainty, 28, 40, 42–46, 48n14, 52, 117n13, 138, 192, 240, 243–4
   myth of, 80, 111
Chabris, Christopher, 62n21
Chalmers, David, 71
chance
   as synonym for probability, 48, 52–54, 115, 117, 119, 211n16
   as synonym for randomness, 79, 214n21, 226, 240
checkers, 92–93
chemist, 103
chemistry, 67, 70, 75, 95–96, 102–3, 214, 228
chess, 46, 92–94
Chomsky, Noam, 107
Christ, 18n10, 182, 225n8. *See also* Jesus.
Christian, 19, 101, 102n51, 119n19, 124, 157n29, 172, 175n7, 176, 181, 206, 224, 229n21, 234n45, 246
Christianity, 19n13, 72n54, 107n69, 109n76, 147n14, 164, 175–76, 197, 211n16, 224n3, 228n19, 229n21, 234n45
chronological snobbery, 166
Churchland, Patricia S., 63n22
Cicero, 51
clarity
   in communication, 31
   God's, 201–22
   in thinking, 82
Clark, Thomas W., 71n52
Clark, William, 141, 152, 154, 191
Clarke, Arthur C., 111–12
classification, 16, 35, 140, 193. *See also* categorization.
clone(s), 94, 124, 212
closed-minded, 205–6, 211, 249
cognition, 70–71
cognitive, xiii, 10, 18, 22, 25, 56, 60, 62–64, 66, 68–70, 81–82, 107, 120, 140, 158, 229
   momentum, 249
   neuroscience, 109
   psychologist, 23, 90
   science, 122
cold fusion, 106

Collins, Francis, 246
Colorado, xiv, 42, 183, 229
  Fourteeners Initiative, 183
  River, 152
Columbus, Christopher, 140, 151, 154, 156
Combs, James O., 250n25
commitment, 121, 149, 151, 154n25, 193, 227
common sense, 85–86, 122
communication, 31, 36, 39, 46, 51n21, 177, 201–2, 221, 246
comparison
  against standards, 96, 98
  of secondary observations, 37
complexity
  computational, 92–94
  God's, 202
  of scientific theories, 176n10
computer, 46, 50n16, 56, 57n5, 81, 83, 93–94, 104–5, 131, 191n37,
  science, 134
  scientist, 159
conjecture, 8, 91, 100, 139. *See also* hypothesis.
consciousness
  as awareness, 89
  and components of faith, 55–63, 66–67, 120, 143, 157n30
  and effort, 86, 116n9, 126, 143, 244
  emergence of, 215
  as explanatory challenge, 70–71, 202
  phenomenal aspects of, 133
  and relation to mind, brain, and soul, 73–75
  studying, 133n46
consensus, 5, 9, 44, 100, 164–65, 178, 225n8, 248
context, 9, 23n26, 31–32, 37, 47, 242
  cerebral, 23, 39, 59–60, 62, 171
  cultural, 11, 121–25
  and evidence, 129–32
  of faith, 39, 113n2
  historical, 132
  linguistic, 9, 15, 41
  religious, xvi, 21, 30, 172, 209, 211n16, 226, 246n18
  scientific, 126, 131–32, 135

sensory, 24, 58
theological, 115n5
Corinth, 181
correlation, 162n41
Cory, Lloyd, 21n19, 223n2
cosmological constant, 103n56
cosmology, 29, 72, 132, 194, 196
creative
  approaches to faith, 139–67
  explorers, 141
  neural network, 64n25
  rationalization, 118
  self-organizing systems, 199
  thinking, 159
creativity
  limits to, 79
  as purpose, 193
  scientific, 79
Crick, Francis, 70n46, 75n65, 197–98, 206n12
Cromwell, Oliver, 13
Crosby, Bing, 29
Crumbo, Kim, 152n18
culture, 11, 121–22, 125, 164, 177, 194, 199, 201, 223
cynicism, 14, 21, 149–50, 153n21
cultural inertia, 121

Damasio, Antonio, 230n27
Darwin, Charles, 5n6, 166, 172n3, 247n19
Darwinism, 102
dating service, 94
Davies, Paul, 72n55
Dawkins, Richard, xiii, 10, 29, 53, 101, 102, 165–67, 204n6
Death Valley, 18
decision tree, 208
deduction, 51n21
deductive reasoning, 35, 86–88, 157n30
Deep Blue, 46, 93–94
deficit. *See* neural deficit.
deism, 201
deity, 5, 38, 72, 79, 120n20, 124, 164, 193, 195, 201, 204, 224–25, 228, 233, 240
  and religious belief, 171–91
  See also God; god(s).

Dennett, Daniel C., 66, 70n46, 133n46, 195
Denver, John, 204, 207
Descartes, Rene, 31, 39, 70n45
Desimone, Robert, 56n2
destiny, 120n20, 243. *See also* fate.
determinism, 119–20, 178, 220n34, 233, 241, 244n13, 245n14
Diamond Creek, 152–53
Diamond, Jared, 11n9
Dilday, Russell H., 38n6, 166n53
dimensions, xiv–xv, 81, 111
Dirty Harry, 80
discovery, 87, 126n30, 226
    exploration and, 141–42, 144–45, 152
    and relation to God's clarity, 219–22
Disraeli, Benjamin, 23n26, 34
Donaldson, Steve, 64n25, 249n24
doubt, 8, 20, 27, 34, 37, 40, 44–45, 47n11, 62, 67, 75, 90, 95–96, 104, 106, 125–26, 146, 152, 159, 233, 236
    absence in Factland, 240
    and the apostle Thomas, 130
    as faith, 113n2
    about God, 202, 213, 218, 249n23
Dowell, William, 94n28
Drummond, Henry, 71n53, 108n73, 126, 171

Edelman, Gerald M., 30n39
edge of chaos, 244n12
Edison, Thomas A., 84
Edsel, 84
Edward III, 84
Egypt, 210, 224
Eiland, Morgan, 4
Einstein, Albert, 31, 36, 38, 79, 102–3, 110, 148, 166n54, 198n23, 226n9
embryology, 166
emergence, 59, 193, 197, 215
emotion, 68n38, 70n44, 75, 133, 145n12, 227–31
environment
    and context for cognitive processing, 57–58, 62, 66, 83, 247–48

    and influence on faith, 120–25
    natural, 11–12, 62–64, 82, 99, 120, 128, 144, 176, 248–49
epistemic circle, 235. *See also* belief cycle.
epistemology, 116n8
error
    of beliefs, 121, 164, 182
    Descartes', 230n27
    due to cessation of thinking, 159
    in experiments, 106n67
    in logic, 89, 206
    in measurement, 95–98, 98n39
    in memory, 89–90
    mixed with truth, 174
    Nietzsche's, 18
    observational, 35, 90
    in speech, 58
    toleration of, 239
    trying to predict God's activity, 225
    typographic, 135
estimation and faith, 12, 68, 91, 117n12
eternal life, 233. *See also* heaven.
Eve, 83
Everett, Hugh, 95n31
evidence, xii, xiv, 5, 41, 43n9, 45, 47, 51n20, 53, 65, 67n34, 70, 85, 96n33, 101–3, 114, 121–22, 135n52
    and characteristics of religion, 194–95, 197–98
    confirming and disconfirming, 136–37, 163, 197, 234
    and exploration, 144–47, 149, 151, 154, 156, 157n29, 160–64, 247–48
    and free will, 220, 244–45
    and God's clarity, 201, 204–8, 210–16, 218–21
    history and, 131–32, 208
    and hypothesis trees, 208, 210–16
    importance of, 125–30, 156, 244
    and importance of context, 129, 131
    and misconceptions about faith, 14–16, 20–21, 30n39, 32
    nature of, 130–33, 205–8, 212–14

evidence (continued)
   and the nature of religious belief, 173, 177–80, 182, 184, 188, 190–91
   quantity of, 135–38
   and traditional views of religious faith, 223–24, 226–32, 234
   and wishful thinking, 118, 144
evolution, 19n14, 29, 38, 66, 104, 128–29, 132, 146, 176–77, 198–99, 247n19, 250n26
   as supposed threat to human significance, 71n53
   as theory or fact, 101–2, 102n51
evolutionary
   biologist, 23, 128n35
   biology, 196
exactness, 95–96, 98n38, 98n39, 100, 103, 106n67
exodus, 224
experience, 8, 15, 17, 22, 43, 51n21, 56, 60, 66, 75, 84, 106, 119, 163, 184–5, 187, 189, 228, 235
   as basis for prediction, 67–68
   religious, 132–33, 202, 219, 227n12, 231
   sensory, 14, 23–27, 32, 35, 43, 121n21, 228
   See also context.
experiment, 17, 25, 30n39, 34, 37, 69n44, 81, 91, 95–96, 100, 103, 106, 110n81, 126, 128, 168, 198, 203, 207, 213n18, 240
explanation, xiv, 12, 116n8, 171
   and God of the gaps, 108–9
   religious, 164, 182, 185–87, 206, 212, 214, 216, 225
   scientific, 70–71, 109–12, 128, 177
exploration
   as basis for genuine freedom, 245
   belief in God's existence as gateway to, 234
explorer(s)
   approach to religious claims, 186n26, 214–15
   attitude of, 153–55
   becoming, 150–51
   behavior of, 155–60
   characterization of, 139–44
   and dangers, 162–65
   fears faced by, 144–50
   instincts, xi
   judgment, 160–62
   and mechanic(s), 139–67
   mindset, 138, 173
   and reasons for believing, 165–67
   and relation between human and divine knowledge, 190–91
   and relation to God, 222
   religious objectives of, 173, 208
   as trustworthy advocates of a religion, 217
   and truth, 191, 245–50
   vision of, 151–53
exponential growth
   in search time, 93
   in sum of human knowledge, 139n1
extrapolation, 75, 95n32, 100n44, 196
eyewitnesses, 212

fact
   versus faith, 64, 84, 160, 239–45
   as supposed certainty, 27, 35n3, 46, 100, 102n51, 105, 107, 127, 132, 192
   versus theory, 101–2
Factland, 239–45
faith
   adaptive nature of, 63–70
   alternative expressions for, 9, 51
   and assumptions. *See* assumptions.
   and attitude. *See* attitude.
   as basis for choice, 8
   and belief formation and assessment, 90
   as belief in the impossible, 17–19
   not binary, 27–29, 48, 67
   blind, xii, 21–22, 47, 117n10, 117n12, 122, 130, 155, 226
   and brains. *See* brain.
   and Calvinism, 119n19
   childlike, 226–27
   complementary, 53–54
   and conflicting beliefs, 228
   and consciousness. *See* consciousness.

SUBJECT INDEX

contagious nature of, 10
and context. See context.
crafting rational, 113–38
creative approaches to, 139–67
definition of, 47–54
and differences between people, 7–12
as disparaged by religious people, 5
dogmatic positions related to, 3–6, 115, 121
and doubt. See doubt.
and emotion, 68n38
emotions and religious, 227–30
and environment. See environment.
and estimation. See estimation and faith.
evaluating, 48–54
and evidence. See evidence.
and explorers. See explorers.
and *fact*. See fact.
and fate, 119–20
and faulty beliefs, 63, 164, 174
and genetics. See genetics.
goal of, 113
and God. See God.
growth of religious, 231–32
and implications of Gödel's theorem, 104–5
and impossibility of existing in isolation, 12
incompatible definitions of, 20
and incongruity between professed and actual beliefs, 22–23
and inevitability of error, 12
and intelligence. See intelligence.
and interpretation. See interpretation.
irrational. See irrational.
and Jesus. See Jesus.
and judgment. See judgment.
and knowledge, 105–11. See also knowledge.
limitations necessitating. See limitations.
not limited to religion, 4–5, 17, 29–31, 33–34, 48. See also religion.
and measurements. See measurement.
mechanistic. See mechanistic.
meta-. See meta-faith.
and memory. See memory.
and miracles. See miracles.
misconceptions about. See misconceptions.
and motivation. See motivation.
and mystery, 71–72
and non-human species, 65–67
and observation. See observation.
of the gaps, 109
operational (functional) depiction of, 63
operative in all human endeavors, 29–31
and paradox. See paradox.
and Pascal's wager. See Pascal's wager.
personal nature of, 11
and prediction. See prediction.
and probability. See probability.
productive ways to think about, xiii
and ranking belief importance, 155–6
rational, xvi, 113–38, 141, 144, 149, 150n16, 154, 160, 162, 165, 180, 182, 205. See also rationality.
and reason, 19–21, 47, 79–167. See also faith, rational.
relational, 227
religious, xiii, xv, 19, 171–91. See also religion.
revealed in common speech, 50, 68
and sacred texts. See sacred texts.
in science, 103n59, 106, 192–200. See also science.
and search. See search.
stable versus precarious, 5–6
substantiated, 224–26
as substitute for sensory experience, 23–27
and things deemed beyond belief, 183–87
and thinking. See thinking.
and traditional religious perspectives, 223–36
and truth seekers, 245–50
variables associated with, 114

faith (continued)
  and wagering. See wager.
  Western, 196
  and what is not yet known, 188–91
fallacy
  logical, 206
  of making the right choice, xii
  of trying to prove via example
    (inductive fallacy), 13, 15, 230
falsifiable, 127–28, 207
Faraday, Michael, 58n11
fate, 91, 167. See also destiny.
Fauconnier, Gilles, 109
fear
  as barrier to exploration, 144–50, 248
  of cynicism, 149–50
  debilitating, 163
  of effort, 147–48
  of embarrassment, 241
  of failure, 148–49
  irrational, 159
  of isolation, 148
  of the known, 145–46
  of the Lord, 189
  of losing cherished beliefs, 5, 137
  of loss of mystery, 72
  of reductionism, 73
  of slippery slopes, 145–47
  of undermining freedom, 244
  of the unknown, 145
feedback, 121, 144, 216n26. See also belief cycle.
feelings, 28, 42, 46, 48, 133, 137, 219, 227–31, 233
Feldman, Michael, 51n20
Ferris, Timothy, 126n30
fideism, scientific, 196
Fitzgerald, Ella, 29
Fleetwood Mac, 220
Flew, Antony, 9, 246–47
forgetfulness, 53, 72, 90, 102. See also memory.
freedom, 29, 69n40, 115, 129n39, 186, 220, 229n20, 243–46. See also free will.
free will, 10, 52
  absent to puppets, xii
  and conscious effort, 244
  no easy answers about, 244
  parroted response about, 53
  and Pascal, 233
  and pre-conscious decisions, 67n34, 69n40
  See also freedom.
Freiberger, Paul, 50n16
Friend, Tad, 41n7
Freud, Sigmund, 127–28
fuzzy logic, 50n16

Galileo, 12n10, 43n9, 84, 126–27, 129, 134, 148, 155
Gall, Franz Josef, 31
Gallup polls, 9
gamble, 117, 127, 240. See also wager.
game tree, 92–94
Gandhi, Mahatma, 148
Garreau, Joel, 81n9
Garside, Rick, 119n17
Gateway Arch, 152
genetic algorithm, 159n35
genetics, 10n7, 56–57, 75, 151, 162, 177
  and faith, 11, 176, 208
Gentiles, 215
geology, 25, 129, 166, 195, 196n10
Gersting, Judith L., 87n23
Giberson, Karl, 5n6, 247n19
Gilovich, Thomas, 9n4, 85–86
Girard, G., 97n37
God, xii, xvi, 7, 10, 27, 30n40, 34, 45, 51–53, 101–2, 104, 116, 124, 126, 129n39, 138, 157n29, 163, 167, 193–94, 198–99, 244,
  clarity of, 201–22
  of the Gaps, 108–11, 185, 214
  and religious belief, 171–91
  and traditional views of faith, 223–36
  and truth seekers, 245–49
  See also deity.
god(s), 9–10, 47n12, 108, 178n14, 184, 193, 199, 219n31, 233
Gödel, Kurt, 104
Gödel's theorem, 104
Goff, Lyn M., 56n2
Goldsmith, Oliver, 55

SUBJECT INDEX

Gould, Stephen J., 175
Grand Canyon, 89, 111, 140, 152
Grand Teton, 183
graph (mathematical), 215n22
graviton, 110n81
gravity, 96, 110, 127, 192
Green, Joel, 74n61
Gregory, Richard, 67
Grogan, Barbara B., 135n50
Gulf of Mexico, 60, 217

Hakel, Milton D., 50n16
Haldane, J.B.S., 128n35
Hamlet, 91
Harris, Sam, 239
Harrison, Peter, 35n3, 70n45, 197n18
Haught, John F., 72, 109, 196n14, 197n18
Hawking, Stephen, 244n9
Hawkins, Jeff, 64n25, 167n55
heaven, 219. *See also* afterlife.
Heilman, K. M., 82n11, 83n14, 175n5
hemispatial neglect, 82–83
Henig, Robin M., 172n3
Herculano-Houzel, Suzana, 80n4
Heisenberg, 198n23
    Uncertainty Principle, 96
heterophenomenology, 133n46
Hinduism, 124, 164
historian, 25, 106, 131–32, 174, 188n32
history, 117, 131–33, 143, 148, 178, 186, 195, 208, 213n18
    of earth, 83
    human, 8, 44, 83, 172, 194, 197, 223
    religious, 38, 132, 212–13, 215, 219, 224
    of science, 31, 67, 102
    personal, 26, 57, 60, 228
    United States, 14
    of universe, 18, 194
Hitchens, Christopher, 129, 246
Hofstadter, Douglas R., 66n29, 67n35, 69, 73n59, 104n61
Horgan, John, 72, 107
hope, 30n39, 70n47, 72, 75, 97, 100, 125, 129, 145, 149, 154, 161, 172, 195, 203–4, 206, 212, 224, 229n21, 234n45, 241, 243n5

and motivation, 117–19, 241
See also wishful thinking.
hopeless, 112, 143, 159, 189
hot hand (in basketball), 85
Hull, Edward, 83n15
Hume, David, 35n3, 67, 162n41, 164n44, 178n14
humility, 29, 174, 188.
Huxley, Thomas, 22–23, 165n48
hypothesis, 29, 33, 37, 54, 58, 60, 67, 69, 70n46, 75n65, 95n31, 100, 116n7, 128, 134, 162, 185, 196, 197n19, 198n26, 206n12, 207–20. *See also* conjecture.
hypothesis tree, 208–17

IAC neural network, 61n18
IBM, 46, 51n20, 93
Idaho, 241
idol(s), xi, 203
    of the mind, 85
idolatry, 198
immanence, 199n29
impossibility, xiii, xiiin4, xv, 14, 17, 19, 21–22, 32, 90, 100, 103, 105–7, 124, 132, 145, 147, 159, 160n38, 166, 178, 184, 186–87, 190n35, 202, 205, 210–12, 220, 228, 234, 239, 242–44, 248–49
Incredible Hulk, 191
induction, 18
    fallacy, 15, 230
inductive reasoning, 13, 35, 44, 86, 88, 157n30
inference, 35–36, 43n9, 61, 64n25, 86–87, 109, 132, 196
intelligence
    artificial, xiii, 70, 71n47, 105
    comparisons, 79, 102
    and naïve views of faith, xv
    as purpose, 193
    and reasoning, 50
    and truth seekers, 246
International Bureau of Weights and Measures, 97n34
interpolation, 100n44
interpretation
    Biblical, 182, 186, 225n8

interpretation (continued)
  of evidence, 129, 162, 179
  of the geological record, 196
  linguistic, 41
  of natural events, 126
  of past events, 102n54
  and religious faith, 178, 180, 198
  and revelation, 38–39
  of sensory experience, 23–24
intractability, 94
irrationality
  accusations of, 246
  and attempts at rationalization, 118, 248
  and behavior, 31
  and beliefs, 21, 146, 230
  and bias, 140
  and defenses of faith, 3–6
  and faith, 121
  and fears, 159
  and presumption about Pascal's wager, 233
  versus rationality, 116n9
  and reasons, 116n7
Isaacson, Walter, 31n41, 103n56, 166n54
Isaiah, 215, 235
Islam, 124, 164, 175n7, 197. See also Muslim.
Israelite, 186, 210, 215, 224
Ives, Joseph, 152–53

Jackendoff, Ray S., 75
Jackson, Reynold G., 183n23
James, William, 69, 201, 227n12, 228
Jastrow, Robert, 161n39
Jaynes, Julian, 70n46
Jefferson, Thomas, 148, 188, 239
Jeffress, Lloyd A., 71n50, 196n12
Jenkins, Mark, 148n15
Jeopardy, xii, 51n20
Jerusalem, 157n29
Jesus, 23, 26–27, 115n5, 128–30, 142n7, 153n21, 160n38, 197, 248n22
  and God's clarity, 206, 210, 213n20, 215
  and the nature of religious belief, 175, 181, 185, 186n20

  and traditional views of faith, 224–27, 230–33
  See also Christ.
Jew, 157n29, 175n7, 185, 197
John, Gospel of, 198, 246
Job, 148, 197
Judaism, 164
judgment, xiii, 79, 140, 153, 225n8, 228
  and faith, 43, 48, 60, 62, 64, 76, 85, 91, 99, 124n27, 139n1, 160–62, 229
Jung, Carl, 116n7

Kahneman, Daniel, 85
Kandel, Eric, 83n13, 83n14
Kasparov, Gary, 93
Kauffman, Stuart, 199, 244n12
Kaufman, Gordon D., 109n74, 124n28
Keizer, Ted, 183
Kelvin, Lord, 99n43, 123n25, 123n26
kilogram, 97–98
King Jr., Martin Luther, 99, 148
Kirshenbaum, Sheril, 146n13
Kitcher, Philip, 101
Kluger, Jeffrey, 176n9
knowledge, 3, 14, 16, 18, 24, 26, 48n15, 53, 55, 70n44, 113, 122, 139, 141, 145, 147–48, 159n34, 165n48, 166, 212, 217, 240, 242
  versus belief, 116n7, 130, 134, 242
  and difficulty of claiming certainty, 79–112
  and feelings, 229
  and freedom, 163n42, 243
  of God, 189, 191, 219
  and how we know things, 35–39
  and religion, 188–91
  and what we know for sure, 39–47
Krauss, Lawrence M., 228n17
Krauthammer, Charles, 94n28
Kuhl, Patricia K., 58
Kuhn, Robert L., xiin3
Kuhn, Thomas S., 103, 106, 122, 142n7, 165, 166n50, 196n16, 197n17, 197n20
Kupfermann, Irving, 83n13, 83n14
Kurtz, Paul, 19n12, 29, 195n6, 196n11

Lakoff, George, 245
Lamarck, Jean-Baptiste, 10
language, 30, 50, 58, 66, 70, 103, 201–2, 244n11
Lanza, Michael, 183n22
Laplace, Pierre-Simon, 185–86, 244
large hadron collider, 135n52
Lashley, Karl, 196
Latour, Bruno, 194n4
learning
　active, 157–59
　as basis for decision, 60
　environmental influence on, 125
　and evidence and free will, 244
　and exploration, 142, 147, 219
　and false beliefs, 122, 126
　and feelings, 230n23
　foundation for, 166
　about God, 174, 189, 190, 202, 207n13, 219–20, 222, 227
　Lamarckian, 10
　and language, 58
　and memory, 56, 81, 107n72
　and parroted responses, 53
　passive, 157–58
　and pool of knowledge, 109
　and prediction, 64
　from secondary observation, 17
　and sensory systems, 81
Lewis, C. S., 19, 107n69, 110n77, 143, 147n14, 166, 175n7, 176, 211n16, 224n3, 228, 229n21
Lewis, Meriwether, 141, 152, 154, 191
Libet, Benjamin, 67n34
light
　illuminating understanding, 18, 21, 102, 153–54, 159, 177, 193, 215, 224
　speed of, 97, 105
limitations, 29, 114, 124, 140, 165
　affecting acquisition of evidence, 138
　and approximate answers to significant questions, 90–95
　as chronological constraints, 83–84
　as inability to assimilate growing body of knowledge, 105–11

inherent in precise disciplines, 104–5
　and logic in brains, 84–90
　and measurements as approximations, 95–100
　necessitating faith, xv, 27, 79–112, 121, 163, 165, 166n49, 232
　and physical constraints of brains, 80–83
　predictive, 243n4
　of science, 193
　and sensory experience, 23–27
　and short-term memory, 87n24
　and theories as approximations, 100–104
Lincoln, Abraham, 17, 148
Lindley, David, 72, 130n42
Lindsey, Hal, 225n8
linguistics, 30, 186
literalness
　and approach to evidence, 206
　and Biblical interpretation, 22, 182, 186n26, 197–98, 219n31, 231
　and science, 198
Lloyd, Seth, 244n9
Locke, John, 11n8, 47n11, 48n14, 51, 56, 101n49
logic, 13, 16, 18–21, 44, 50n16, 71n47, 84–89, 100, 104, 129–30, 134–35, 161n40, 206, 214, 227n12, 228, 230
Lord, 21, 79, 181–82, 188–89
Lorentz, Edward N., 62n19
Lotto, R. Beau, 57n7
Love, Mike, 228n13
Lucretius, 195
Luger, George F., 188
Luther, Martin, 83

MacPhail, Euan, 66n32
magic, xii, 112
Mallory, George, 110, 147
Manhattan Island, 153
Maritain, Jacques, 107, 134–35
Martini's Law, 82
Marx, Karl, 127
Maslow's hierarchy of needs, 148
Mather, Increase, 21–22

mathematician, 58n11, 71, 104, 215
mathematics, 48, 81, 87n23, 96, 100, 104–5, 109, 134, 185n24, 196, 221
McClay, Wilfred M., 144n10
McClelland, James L., 61n18
McCulloch, Warren S., 70n47
McGhee, George, 102
McGilchrist, Iain, 228n16
McGrath, Alister, xiiin4
McNamara, Thomas E., 60, 62n20, 63n23, 67n34, 121n21, 196
McNeil, Daniel, 50n16
measurement, 54, 94, 222
    of computational resources, 92–93
    of faith, 117
    geometric, 46
    as limitation, 80, 95–100
    scientific, 41, 196
    and strength of evidence, 47n13
Mecca, 157n29
mechanic(s)
    approach to religious beliefs, 190
    and arrogant explorers, 164
    characterization of, 139–44
    explorer(s) and, 139–67
    and search, 156
    and vision, 152
mechanistic
    approach to faith, 143
    theories, 102
    universe, 243
Meldahl, Keith H., 196n10
memory
    anomalies, 89–90, 158
    as basis for faith, 57, 60
    foundational aspects of, 56–57, 75
    limitations, 81, 87n24, 93
    types, 56
Mencken, H. L., 19–21, 23, 150n16, 210, 236
Meno, 235
mental image, 42, 82, 89, 124, 140–41
meta-faith, 4, 13, 115n3
metallurgy, 159
metaphysics, 44n10, 76, 244
meter, 97
Michelangelo, 153

Mill, John Stuart, 159
Miller, George A., 81n6
mind, xv, 37, 51, 70–71, 73, 75, 85, 101, 107n70, 125, 130, 137, 143, 148, 164, 173, 196, 204, 210–11, 230, 235, 245–46, 249
mind-body problem, 70–76
mindset, xv, 3, 103, 129, 138, 141, 163, 188, 210
Miner, Horace, 124n27
Minsky, Marvin, xiii, 57n6, 74n60
miracle, 26–27, 43, 50, 110n81, 177–80, 184–86, 208, 210–11, 216, 219–20, 224
misconceptions
    about faith, 3–6, 13–32, 47, 51n21, 128n36, 163
    that faith means belief in the impossible, 17–19
    that faith is binary, 27–29
    that faith is blind, 21–23
    about faith and evidence, 14–17
    that faith and reason are incompatible, 19–21
    that faith is just a religious phenomenon, 29–31
    that faith is a substitute for sensory experience, 23–27
Mlodinow, Leonard, 4n4, 58n11, 85n17, 91n26, 129n38
Mohammed, 210–11, 213
Montaigne, Michel de, 124n27, 130
Mooney, Chris, 146n13
Moravec, Hans, 71n47, 81n5, 104n63
Morowitz, Harold J., 48n15, 235
Morris, Simon Conway, 102n53, 196n14
Mother Teresa, 148
motivation
    factors contributing to, 148
    and faith, 31, 69
    and feelings, 230
    and hope, 117–19, 241
    and rationality, 149
    to seek insight/truth, 190, 241
Mount Everest, 110, 205
Mount Moran, 25–26
Muslim, 157n29, 213. *See also* Islam.
mystery, 70–72, 109, 127, 185–86

SUBJECT INDEX

Naam, Ramez, 81n9
Napoleon, Bonaparte, 185
NASA, 18
National Center for Science Education, 196n11
National Geographic, 134
National Public Radio, 204n6
National Weather Service, 117n13
natural selection, 101
naturalism, 196n14
nature (biological/physical), xiii, 10, 101n48, 102, 126–27, 166–67, 171–72, 177–80, 186–88, 195, 196n14, 199n29, 221
neglect. See hemispatial neglect.
neural deficit, 82
neurologist, 82
neurons, 59, 61, 69, 75, 80–81, 111, 120, 227
neuroscience, 53, 57, 75, 109, 230
neuroscientist, 57, 70
Newton, Isaac, 79, 102–3, 110, 148, 156, 243
Nietzsche, Friedrich, 18, 202
NIST, 98n38
non-computability, 71. See also Gödel's theorem.
non-equilibrium thermodynamics, 103
North America, 18, 241
number theory, 104n61
Numbers, Ronald L., 12n10

objections, 150n16, 219n29
    religious, 223–36
observation
    to obtain evidence, 130–31
    primary, secondary, tertiary, 27, 35–39, 43–44, 51n21, 55, 58, 60–62, 66–68, 84, 86, 89–90, 106, 127–28, 178, 197–98, 245
    probabilistic role in faith, 48, 55
    unremarkable, 45
    See also perception.
Old Testament, 215, 224
old wives' tales, 86
Olympian, 148, 184
omnipotence, 192, 225
omnipresence, 84
omniscience, 79, 157, 192, 225, 240–41, 243
omnitemporal, 84
open-minded, 211
optimality, 48, 92, 94
ornithologist, 84
Ortenburger, Leigh N., 183n23

paleontologist, 24, 101–2
Panek, Richard, 3, 27n31, 43n9
panspermia, 198
paradigm
    cultural, 122
    in non-scientific domains, 122n22
    religious, 104n60, 122, 197n20
    scientific, 28, 103, 122, 142n7, 166n50, 196–97
paradox
    of increasing knowledge, 105
    of knowledge and ignorance, 240
    of mechanic beginnings, 143
    of omniscience and freedom, 243
    of possibilities and limitations related to faith in God, 232
    of searching, xi
    and a world of certainty, 240
parallel universe, 94, 95n31
parent, ix, 46, 100, 125, 154–55, 226–27, 229, 232
Parker-Pope, Tara, 158n31
Pascal's wager, 51, 53, 232–36, 249
Paul, Saint, 22, 36, 57, 181–82, 219n31, 222n36, 224, 226–27, 231, 235
Peacocke, Arthur, 29, 105n64, 199n29
Penrose, Roger, 59n13, 71n48, 104
perception(s), xi, 26, 43, 57–58, 60, 62, 67, 91, 162, 173, 185, 208, 216, 220n34, 236. See also, observation.
perceptual magnet effect, 58
perfection, 160, 162
Peterson, Gregory R., 119n14, 244n11
Pew polls, 9
philosopher, 18, 35n3, 51, 60, 65–67, 70–71, 86, 101, 103, 107, 126–27, 134, 141, 195

philosophy, 55, 126, 134, 156, 173, 245, 249
physician, 9, 24, 99, 118, 129, 142, 180, 191
physicist, 30, 72, 74, 96, 100, 106n66, 107n70, 135, 195, 207, 210, 244n13
physics, xv, 16, 43n9, 53, 61, 67, 72, 75, 96, 107, 111n83, 126, 134–35, 192, 196, 205, 214, 216, 221
physiologist, 24
physiology, 56, 59, 148
Pilate, Pontius, 153n21
Pillemer, David B., 56n4
Pinker, Steven, 64n25, 244n11
Pirsig, Robert M., 116n9
Pitts, Walter, 70n47
Planck, Max, 30, 197
Plantinga, Alvin, 65–66
Plato, 70n45, 97, 113n1, 235n49
Poe, Edgar Allan, 84
Polanyi, Michael, 33, 154n25, 229n20
Polkinghorne, John, 29
pollster, 100. *See also* Gallup; Pew.
Pope, Alexander, 102
Popper, Karl, 60, 127–28
Powell, Baden, 123n25
Powell, John Wesley, 140, 152
prayer, 162n41, 194–95, 207, 214n21, 216, 243n6
preacher, 182
preaching, 195, 250n25
pre-Cambrian rabbit, 128n35
preconceptions. *See* assumptions.
prediction
    and faith and adaptive advantage, 64–69, 118
    mechanical, 141
    problems for, 63, 67, 110, 192, 243n3, 243n4, 244
    and religious faith, 225–26
    scientific, 64n25, 67, 102, 110–11, 128, 192
premise, 54, 70, 88, 100, 104, 134, 197, 219. *See also* assumption.
President, 40, 100, 130
pride, 6, 31, 163, 173n4, 175n6. *See also* arrogance.

Prigatano, George P., 82n11, 83n14, 175n5
probability, 20, 91, 173, 210, 247
    and adaptive nature of faith, 63–65
    and attention deficits, 62–63
    automatically assigned by brains, 55–56, 120
    and categorization, 132
    conditional, 215n23
    and contextual influences, 59–62
    and Factland, 239, 241–42
    and free will, 243–44
    and evidence, 118, 129–30, 136–8
    and faith as probabilistic understanding, 47–54, 68–69, 76, 83, 112, 113n2, 114–15, 116n7, 243n4
    and hope, 117n11, 117n12, 118, 119n16
    and neuron behavior, 59
    and perception, 57–58
    and relation to rules in logic, 87n24
    and religious beliefs, 178, 198, 210n15, 212, 214n21, 215, 218, 223, 226, 232, 234
    See also belief probability.
progress, 120, 122, 143n10
    in exploration, xii, 152–53, 163
    in forming accurate beliefs, 54, 84, 121, 143–44, 160
    intellectual, 109
    religious, 174, 207n13, 219, 231
Protestant Reformation, 122n23, 197
Proust, Marcel, 152, 159
pruning (search spaces), 68n39, 92
Proverbs (book of), 65n26, 190. *See also* Scripture Index.
Psalmist, 188–9
psychologist, 23–24, 60, 70, 85, 90, 124n27, 129n38, 196
psychology, 56, 75, 159, 214, 230
Ptolemy, 102
puppets (and free will), xii
Purkiser, W. T., 69n44
purpose, 10, 31, 156, 171, 173, 177, 190n35, 193, 196n14, 214, 216, 247n19
Purves, Dale, 57n7

quantum
    effects in brains, 59
    physicist, 100, 221, 244n13
    theory. *See* theory, quantum.
    tunneling, 112, 191
question(s)
    and approximation, 80, 90–95
    big, xii, 9, 19, 111, 115, 140, 162, 245, 246n18, 247
    about evidence, 125, 135–36, 162
    and exploration, 143–44, 147, 151, 158, 161–62, 165–66
    in Factland, 240–42
    about hope, 119n14
    and knowledge acquisition, 36, 178n13
    and probability, 48, 118
    religious, 26, 177–78, 182, 186, 188, 190, 204, 211–13, 218, 224n4, 227
    about science, 192, 196n11
    scientific, 59, 66, 70, 85, 109, 111n83
    of trust, xiii, 117

Rackham, H., 51n18
Rader, Lyell, 21, 23, 223n2
Radowitz, John von, 135n52
Rampino, Michael, 161n39
randomness, 178, 214n21, 243n7
rationality, 116n7, 116n9, 129, 143, 147, 149, 151, 161, 164, 165n46, 231, 248
    emotions, faith, and, 228–30
    See also faith, and reason. *See also* faith, rational.
rationalization, 118, 127, 154n23, 163–64
reason
    deductive. *See* deductive reasoning.
    and faith. *See* faith and reason.
    inductive. *See* inductive reasoning.
Red Sea, 211
relativism, 125, 164
relativity, theory of. *See* theory of relativity.
religion, xi, xiii, xvi, 4–5, 15, 19, 23, 28–31, 34, 48, 56, 103, 104n60, 112, 118, 122, 132–33, 155, 171, 173–78, 180–82, 185, 188, 191, 246–47
    evaluating claims of, 208–17
    and God's clarity, 201–22
    and nature of religious belief, 171–91
    science as, 192–200
    and traditional perspectives on faith, 223–36
Reppert, Victor, 196
resurrection, 102n51, 224
revelation, 38–39, 156, 165–67, 180–81, 190, 195, 208, 211, 219n29
Revelation of St. John, 225n8
Roach, Gerry, xiii
Robbins, Ray Frank, 225n8
Robbins, Trevor, 59n14
Robinson, Marilynne, 75n65, 132n45, 223
Rocky Mountains, xi, 41, 84, 183.
Roco, Mihail C., 81n9
Roediger, Henry L., 56n2
Rogers, Roy, 29
Rolston, Holmes, 30n40
Ross, Hugh, 69n44, 133n45
Rowell, Galen, 58, 89
Royal Society, 165
Ruffin, Steven A., 8n2
rule(s)
    in logic, 87
    spelling, 56
Rumelhart, David E., 61n18

Sacajawea, 154
Sacks, Oliver, 75n64
sacred text, 177, 180–82, 194. *See also* scripture.
Sagan, Carl, 100–101
sample space, 92. *See also* search space.
Sartre, Jean-Paul, 51
Saxe, John Godfrey, 135n54
Schacter, Daniel L., 23–24, 82n11, 83n14, 175n5
Schneider, Susan, 75n63
scripture, 39n6, 180, 190, 209, 215, 217n26, 226, 234. *See also* sacred text.

Schrödinger, Erin, 107n70
Schults, W., 82n10
science, xv, xvi, 9, 27–31, 41, 51n20, 56, 60, 67, 72, 100–103, 104n60, 106–7, 109, 118, 122, 126–28, 130n42, 131–35, 154, 161, 165, 175–77, 187, 207, 216, 224n7, 225, 226n9, 250n26
    as religion, 192–200
science and religion, 29, 118, 122n23, 175, 194n4, 198–99
scientist, 24, 28, 31, 34, 37, 51n20, 71, 85, 99, 101, 103, 106, 131–32, 142, 159, 165–66, 176, 195–99, 207, 235, 240, 244
Scott, Eugenie, 196n11
SCUBA, 82, 217
search
    for answers, xii, 189
    belief about efficacy of, 153
    for commonalities, 42
    for counterexamples, 129n38
    and discovery, 142–43
    for evidence, 156–57, 160, 248n22
    as exploration versus rationalization, 154n23
    and fear of effort, 147
    foundation for, 157
    as fundamental human behavior, 155–56
    and hypothesis trees, 210
    to increase probabilistic accuracy of faith, 157, 226
    and judgment, 160
    and learning about God, 190
    premature end of, 163–64
    and previously charted territory, 154
    space(s), 68n39, 92–94
    for truth, xiv, 115, 118, 125, 145, 149, 165, 175n6, 246–50
Searle, John R., 71
second, definition of, 97
sensory experience. See experience, sensory.
sensory systems, 36, 56n2, 57, 63, 81
Shakespeare, William, 39n6, 98n40, 153, 166n53, 167, 202
Shannon, Claude E., 93–94

Shermer, Michael, 51n20, 163n42, 204, 234–35, 246, 249n23
Shylock, 98
Sierra (Nevada Mountains), 82
Simon, Daniel, 62n21
Simonton, Dean Keith, 79n2, 154n24
Simpson, Ray, H., 50n16
simulated annealing, 159n35
skepticism, 101, 132, 139, 149–50, 166n48, 167, 204, 229
slippery slope, 186n26
    fear of, 145–47
Smith, Christian, 177n11
Smolin, Lee, 195
sociobiologist, 146
sociobiology, 195
sociology, 159
Socrates, 113, 156, 235
Soon, Chun Siong, 67n34
soul, xii, 10, 31, 70, 73–75, 230
speech scientist, 58
Spielberg, Nathan, 110n78, 110n80, 111n83
Spirit (of God), 182, 203
Spitzer, Robert J., 141
SpongeBob, 191
standard, 81, 96–100, 107, 131, 132n45, 135, 160, 198n26, 206, 226
Stannard, Russell, 72n57
Stapp, Henry, 59n13
Steele, Brian, 188n32
Stravinsky, Igor, 244n11
straw man, 19, 74n60, 176, 234n45
supernatural, 179, 186–87, 199n29
Swan, K., 82n10
Swinburne, Richard, 178n14, 187n30
syllogism (and syllogistic reasoning), 87

Taleb, Nassim Nicholas, 198n26
Taunton, Larry, 246
Taylor, Barbara, 115
teacher, 18n11, 91, 126, 142, 154, 182, 231
teaching, 122, 139, 142, 229n21
technology, 4, 94, 111–12, 195
Teilhard de Chardin, Pierre, 43n9, 71n53, 101–2, 192
Tennyson, Alfred Lord, 38

SUBJECT INDEX

Teton range, 25, 183.
thalidomide, 115
Thatcher, Bryce, 183
theism, 187n30, 208
    monotheism, 197
    panentheism, 199n29
    pantheism, 199
theist, 72, 173, 180, 182, 194, 201, 203–7, 217–20, 224–25, 232, 235, 245
    monotheist, 28
    polytheist, 193, 197
theologian, 29, 72, 126, 138, 199n29, 221, 233n39
theology, 53, 57, 115n5, 119n19, 129n39, 166, 171, 174, 177, 187, 193
theorem, 100, 104, 104n61
theory
    of automata, 71n50
    of brain, 63n22
    cosmological, 132, 134
    of consciousness, 59n13
    electromagnetic, 177
    end of, 162n41
    evolutionary, 29, 101–2, 104, 128, 177, 250n26
    falsification of, 127–28
    Freud's psychoanalytic, 127
    geological, 196n10
    of God, 104
    grand unified, 205
    of gravitons, 110
    of history, religion, etc., 103
    of intelligence, 71n47
    judged by explorers, 156
    limited, 69n44, 133n45
    Marx's economic, 127
    number, 104n61
    perceptual magnet, 58
    of planetary motion, 185
    quantum, 20, 96, 178, 243n7
    of relativity, 36, 103, 110, 167n54, 226n9
    of schizophrenia, 124n27
    scientific, 5, 19, 38, 72, 80, 100–104, 111, 128, 141–42, 166, 176n10, 195–98, 214, 246
    of social conflict, 104
    of symbolic processing and language, 70
thermodynamics
    laws of, 192–93
    non-equilibrium, 103
thinking
    and active learning, 157–59, 189
    algorithmic nature of, 104
    Aristotle's influence on, 127
    and assumptions, 136
    and context/environment, 59–60, 82, 124
    creative, 159
    critical, xiii, 13, 158
    and cultural relativism, 125
    and deceptions, 127–28, 159, 164, 174, 176, 228, 250
    and determinism, 120n20
    effort of, 147
    and evidence, 130–31
    and fact, 101
    and faith, xiii–xvi, 4–5, 9, 14–15, 18, 22, 27–29, 34, 47n11, 52, 55, 63, 67–69, 129n39, 140, 155, 172, 218, 224, 241
    and foregone conclusions, 128
    and hypothesis trees, 208–17
    intentional, 158
    and knowledge acquisition, 35, 37–38
    and memory, 56
    and morals, 51n20
    muddled, 85–87, 90
    and neurons, 59
    Nietzsche's, 202
    and probability, 52, 54, 114
    and rational faith, 127
    versus reality, 191
    religious, 118, 138, 176, 180, 186–87, 189–90, 194, 198, 202, 206–7, 218n28, 223, 230n25, 231–32
    scientific, 184, 197
    wishful, xii, 22, 69, 116–19, 127, 130, 137, 144, 163, 193, 211, 215, 223
Thomas (apostle), 130
Thomas, Cal, 129–30, 129n39

Thompson, William. *See* Kelvin, Lord.
thought trajectories, 60–61
Tillich, Paul, 233n39
time machine, 213
time of day, 98n38
tooth fairy, 203
tractability. *See* intractable.
tradition, xvi, 38, 165–66, 177, 194, 196–99, 208, 220n34
transcendence, 199n29
Trivial Pursuit, 80
truth
  and conflicting arguments, 4
  and consensus, 164
  and consistency, 16
  content of beliefs, 28, 47n11, 65, 116, 116n8
  versus convenience, 248
  dilemma of knowing, 160–62
  and error, 144, 174, 247n19, 249–50
  and evidence, 125–26, 131–32, 156
  and explorers, 152–54, 156–57, 160, 191
  and fear, 145, 239
  and feelings, 230n23
  freeing effect of, 115n5, 153
  and inference, 64
  misunderstood, 201
  and question of desire, 115, 191n38, 232
  religious, 173–75, 211–13, 217
  science and, 195, 197n18
  scientific paradigm as, 103
  searching for, xiii–xiv, 115, 118, 125, 144n11, 175n6
  seekers, xvi, 165, 173, 245–50
  skeptical versus cynical approaches to, 149–50
  theorem as, 104
  value of pursuing, 146, 153, 160
Turing, Alan M., 70n47, 104n63
Turner, Mark, 109
Tversky, Amos, 85

uncertainty, 21n18, 23n26, 40, 85n18, 96
Upheaval Dome, 195, 196n11

Utah, 5, 153, 195

vision
  of explorers, 151–53
  as foresight, 242
  as goal, 80, 135, 239n1, 245
  as insight, xiv, 22, 185, 240
  ocular, 35n4, 57–58, 63n22, 81, 97
  religious, 171, 175, 186–87
Von Neumann, John, 71
Vos Savant, Marilyn, 228, 233–35

wager
  as manifestation of faith, 51
  Pascal's, 51, 53, 232–36, 249
  See also gamble.
Walker, Evan Harris, 59n13
Walsh, P. G, 51n18
Ward, Keith, 75, 138, 178n14, 196, 244n13
Watson (IBM), 51n20
Weil, Simone, 157
Weinberg, Gerald M., 29–30
Western hemisphere, 81, 241
Western thought, 188–89
Wilder, Thornton, 28
William of Conches, 7
Wilson, Edward O., 11n9, 194, 246
Wilson, Brian, 228n13
wishful thinking, xii, 9, 22, 54, 69, 116–19, 127, 130, 137, 144, 163, 193, 211, 215, 223. *See also* hope.
Wolfram, Stephen, 244n9
Woolgar, Steve, 194n4
worship
  at altar of cultural relativism, 125
  characteristics of, 194–95, 200
  and response to miracles of Jesus, 27
  self-perceptions, xi
Wyoming, 25, 133, 183

Yerxa, Donald A., 143n10
Young, Simon, 195

Zacharias, Ravi, 230n23
Zadeh, Lofti A., 50n16
zombie, 120

Scripture Index

Psalms

131:1	188
14:1	23n27
39:4	79

Proverbs

2:3-6	189, 190
6:6-8	65n26
16:18	173n4

Isaiah

49:6	215

Jeremiah

5:21	164n43
29:13	191n37

Daniel

5:27	99n41

Amos

6:1a	229

Matthew

5:14	215
5:45	160n38
7:7-8	190, 231n28
13:44-46	236n55
16:3b	248n22
17	186n26
17:20	69n43, 231n33
18:3	232n36
18:8-9	23n24
19:14	226, 232n36

Mark

3:1-6	26n30, 128n36, 213n20
9:37	232n36
10:2-12	181n17
10:15	226, 232n36
12:30	230n24

Luke

5:36-39	142n7
7:50	233
9:48	232n36
16:15	236n55
16:31	206
17:5-6	234n46
17:6	69n43, 231
18:8	224n4
18:17	226, 232n36

John

1:18a	198n24
4:12a	198n24
4:48	206, 213n20
6	186n25
8:12	215
8:31–32	247n21
8:32	153n21
9:5	215
9:41	175n5
11:1–53	213n20
14:11	224
18:38	153n21
20:24–29	130n40

Romans

1:17	224

1 Corinthians

1:25	222n36, 232n36
1:26	226
2:6–10	231n31
3:2	231
7	22n23
7:10–13	181
7:29	182n19
7:31	182n19
7:40	182n20
9:19	244n11
13:11	231
13:12	5n7, 219n31

2 Corinthians

5:7	57

1 Timothy

4:8	236n55

Hebrews

11:1	96n33, 224
11:6	234

James

1:6b	249n23

1 Peter

3:15	229n21, 234n45

www.ingramcontent.com/pod-product-compliance
Lightning Source LLC
Chambersburg PA
CBHW032054220426
43664CB00008B/991